Digital Twin Adoption and BIM-GIS Implementation

The aim of this edited book volume is to present new concepts, and applications of Digital Twin and relevant tools in the built environment context. The book identifies key organisational factors that influence the adoption of technology within the architectural service industry, setting the stage for a deeper understanding of the shift towards digital methodologies.

The book introduces the Digital Twin Readiness Level framework, a modified metric system with ten levels of risk-based maturity and an empirical development of a Digital Twin Adoption Model. Furthermore, the text ventures into immersive virtual environments and their applications, showcasing innovative practices that enhance learning and operational efficiency.

Additionally, the book examines the integration of Building Information Modelling (BIM) and Geographic Information Systems (GIS), revealing new workflows for creating 3D models of cities. A focus on Australia's government-funded infrastructure projects offers insights into the values and acceptance criteria for these technologies, providing a comprehensive overview of their practical implications and future potential. This book is essential reading for practitioners, engineers, and managers involved in any phase of the built environment from design to operation and other architectural, engineering, and construction (AEC) stakeholders who are a part of digital transformations, as well as researchers, educators, and students interested in the field.

Samad M. E. Sepasgozar is a top 2% researcher in the world computed by the Standard University and the first top researcher (field leader in Architecture) in Australia found by *The Australian*. He is an academic editor (editorial board member) at Scientific Reports Nature and Associate Editor of Architectural Engineering ASCE (Decile 1). He is also a top 1% reviewer globally, an editor, editorial board member, or reviewer for 60 leading high-ranked journals, and a lead assessor for national research projects on innovation and smart technologies. He published over 200 peer-reviewed articles and received various international and national awards annually. He is an Associate Professor at the University of New South Wales (top 19 globally), Sydney, Australia.

Sara Shirowzhan is a Senior Lecturer and the Co-convenor of smart cities and infrastructure cluster in the School of Built Environment at the University of New South Wales (UNSW), Sydney, Australia. Her areas of research in technologies relevant to built-environment informatics include sensing technologies, advanced GIS, BIM, Digital Twins, and Artificial Intelligence. She teaches and supervises students on Construction Informatics, City Analytics, GIS, and BIM-relevant topics at undergrad and postgrad levels. She is currently an Editorial Board member of the *Journal of Sustainability* and *Advances in Civil Engineering*. She is also a Topic Board member of the *ISPRS International Journal of Geo-Information*. She completed her PhD in Geomatics Engineering from the School of Civil and Environmental Engineering at UNSW.

Advanced Digital Technologies for the Built Environment
Series Editor: Samad M. E. Sepasgozar

Digital Twin Adoption and BIM-GIS Implementation
Edited by Samad M. E. Sepasgozar and Sara Shirowzhan

Digital Twin Adoption and BIM-GIS Implementation

Edited by
Samad M. E. Sepasgozar and
Sara Shirowzhan

Routledge
Taylor & Francis Group
LONDON AND NEW YORK

First published 2025
by Routledge
4 Park Square, Milton Park, Abingdon, Oxon OX14 4RN

and by Routledge
605 Third Avenue, New York, NY 10158

Routledge is an imprint of the Taylor & Francis Group, an informa business

British Library Cataloguing-in-Publication Data
A catalogue record for this book is available from the British Library

ISBN: 978-1-032-56933-8 (hbk)
ISBN: 978-1-032-82931-9 (pbk)
ISBN: 978-1-003-50700-0 (ebk)

DOI: 10.1201/9781003507000

Typeset in Times New Roman
by codeMantra

Contents

Contributors

Ayaz Ahmad Khan
Australian Research Centre for Interactive and Virtual Environments (IVE), UniSA Creative
University of South Australia
Adelaide, Australia

Hesham Algassim
School of Built Environment
University of New South Wales
Sydney, Australia

Lee Butler
School of Built Environment
University of New South Wales
Sydney, Australia

Mahsa Chizfahm
School of Architecture and Environmental Design
Iran University of Science and Technology (IUST)
Tehran, Iran

Steven Davis
School of Civil and Environmental Engineering
University of New South Wales
Sydney, Australia

David Edwards
School of Engineering and the Built Environment
Birmingham City University
United Kingdom
and
Faculty of Engineering and the Built Environment
University of Johannesburg
South Africa

Juan Sebastian Garzon Romero
School of Built Environment
University of New South Wales
Sydney, Australia

Ning Gu
UniSA Creative, IVE: Australian Research Centre for Interactive and Virtual Environments
University of South Australia
Adelaide, Australia

Ruiyu Liang
School of Minerals and Energy Resource Engineering
University of New South Wales
Sydney, Australia

Alberto De Marco
College of Engineering Management
Politecnico di Torino
Torino, Italy

Subin Mecheril Binoy
University of New South Wales
Sydney, Australia

Marco Mura
College of Engineering Management
Politecnico di Torino
Torino, Italy

Peyman Najafi
Smart Architectural Technologies, The Built Environment Department
Eindhoven University of Technology
Eindhoven, The Netherlands

Joung Oh
School of Minerals and Energy Resource Engineering
University of New South Wales
Sydney, Australia

Michael J. Ostwald
The School of Built Environment
University of New South Wales
Sydney, Australia

Christopher Pettit
City Futures Research Centre
University of New South Wales
Sydney, Australia

Samad M. E. Sepasgozar
School of Built Environment
University of New South Wales
Sydney, Australia

Sara Shirowzhan
School of Built Environment
University of New South Wales
Sydney, Australia
and
City Futures Research Centre
University of New South Wales
Sydney, Australia

Ali Soltani
FHMRI Injury Studies
Bedford Park, South Australia
and
UniSA Business, University of South Australia
Adelaide, Australia

Chengguo Zhang
School of Minerals and Energy Resource Engineering
University of New South Wales
Sydney, Australia

1 Introduction

Artificial intelligence, digital twin transformation, and enabling technologies

Samad M. E. Sepasgozar

1.1 From basic to self-generation digital twins under a meshed network

A digital twin is a virtual representation of a physical entity which offers the possibility to evolve in real-time that can be helpful for understanding processes, learning, reasoning, predicting incidents, and informing decisions. There are notable expectations and evolutions in the concept of digital twins, as discussed by Sepasgozar (2020). However, advances in artificial intelligence (AI) can significantly improve the quality of digital twins and consequently increase the value of outcomes with fewer human interventions.

Digital twins can be further advanced to be *self-generational* and *self-maintained* (Saracco 2021), where the digital version of the physical entity can be autonomously created, updated, and maintained by itself. The digital practice can be further advanced at a larger scale, such as in a city or the entire universe, where numerous *interconnected networks of twins can create meshed networks of digital twins* to share digital information with stakeholders. This allows selected stakeholders to use it, update it, or connect it to their multiple interconnected virtual representations of physical entities, systems, and behaviours.

The continued advances of digital twins and enabling tools open up possibilities to revolutionise project-site construction processes. The integration of blockchain technology and the implications of recent AI practices will improve transparency, decentralisation of data storage, data immutability, secure peer-to-peer communication and offer processed data informing decisions. As digital twin technology is evolving continuously, the technology uptake rate will constantly increase. This holds immense potential for transforming the construction industry and associated sectors. Consequently, the key task to be undertaken is to explore barriers, adoption factors, readiness measures, potential applications, and solutions to current problems in the context of construction.

Figure 1.1 shows the result of a search using Google Trends. Using digital twin as the keywords in the search resulted in an ascending trend (top) in a short period of time. However, using both the keywords AI and digital twin (bottom) shows a significant increase in the popularity of AI, whereas the digital twin does not show a significant increase compared to AI. The *y*-axis of the figures shows the relative

DOI: 10.1201/9781003507000-1

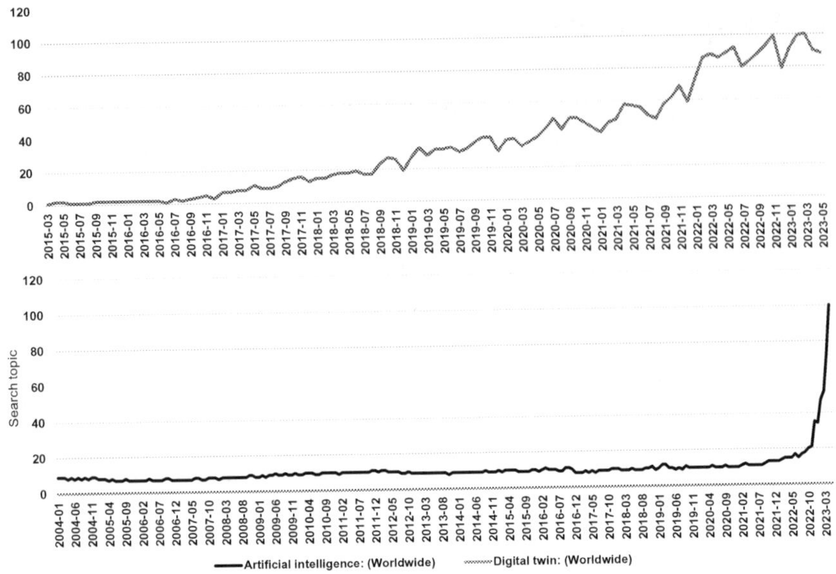

Figure 1.1 Search interest trends. The digital twin trend is above and the digital twin and AI graph is below.

Note: The *y*-axis on the chart shows the relative search interest or activity to the highest point recorded for that time worldwide. When the topic has a value of 100, it means it has reached peak popularity. If the value is 50, it indicates that the term is half as popular as the peak.

Source: Generated by Google Trends.

search activity to the highest point recorded in the time period. While the analysis of popularity needs more accurate data for accurate comparisons, the significant increase in AI applications compared to digital twins as a whole, is quite clear during the chosen time periods.

The digital twin has increasingly become a feasible solution to issues and problems rather than just a distant vision, due to the advances in various tools and technologies, including those listed in Table 1.1. These technologies have paved the way for the development, advancement, and implementation of digital twins, allowing them to become a practical and valuable asset for the construction industry.

1.2 Shifting the concept of digital twin over the years

The functionality of digital twins has transformed from being a static representation of physical entities to a dynamic, data-driven tool. This evolution enables the optimisation of multi-modal complex systems in real time. Going forward, the expectation is to have further advances towards self-generating twins within an interconnected network of numerous twins. As digital technologies and AI continue to advance, the capability of digital twins will continue to evolve further in

Table 1.1 Evolution of enabling technologies and their value for digital twin development

Selected enabling technologies	*History of early versions*	*Examples and the value of the enabling technology*
• Early computer-aided design (CAD) and Simulation • 3D modelling • Photogrammetry (to create a 3D model of an object)	After the 1960s, the development of basic drawing programs, CAD, and simulation tools laid the foundation for 2D and 3D visualisations. Advanced versions of these tools can be used as a foundation for digital twins.	AutoCAD, SolidWorks, Computer Aided Three-Dimensional Interactive Application (CATIA), Creo, Maya, 3ds Max Design, Sketchup, Onshape, Blender and Fusion 360.These tools are used to create 3D modelling, animation, rendering, and behaviour simulations.
• Virtual prototyping • Virtual reality (VR), augmented reality (AR), mixed reality semi-immersive and immersive VR (e.g., through the Window VR, Mirror World VR, Waldo World VR, Chamber World VR, Cyberspace VR) • Holography • Motion capture	After the 1980s, virtual prototyping became a popular application of creating virtual environments. These applications can be used for creating a virtual environment for digital twins and optimise the performance of physical items or products during the design stage, before they were built.	Unity, Amazon Sumerian, Google VR for Everyone, Unreal Engine, ShapesXR, SparkAR, Spline, Tvori, Gravity Sketch and Arkio.The tools enhance visualisation and communication, allow remote collaboration by providing shared virtual spaces, and simulate various scenarios.
• Internet of Things (IoT), sensors, actuators, edge devices, gateways, wearables, and smart systems • Big data • LiDAR scanning • Structured light scanning	After the 2000s, the emergence of the IoT and big data analytics created new opportunities for creating smart systems which can be used for digital twin development. The concept of "mirror world" was introduced, similar to a digital twin, and Nasa developed initial versions of a digital twin (Sepasgozar 2020).	Sensing technologies and other connected devices could be used to gather real-time data about physical objects and their behaviour and transmit the data to the cloud or other centralised systems; this could then be used to create more accurate digital twins.

(*Continued*)

Table 1.1 (Continued)

Selected enabling technologies	*History of early versions*	*Examples and the value of the enabling technology*
Industry 4.0/Construction 4.0(Smith and Sepasgozar 2022): this concept increased the interconnectivity between tools and processes and it emphasises the data analytics in the industry, including smart cities, the construction and building sectors.	After the 2010s, the industry focused on the integration of advanced digital technologies to improve industrial processes.	Various tools and technologies and their integration, are used to improve processes such as IoT, AI, robotics, 3D printing cybersecurity, blockchain, edge computing, and advanced wireless technologies (Kerin and Pham 2019; Sepasgozar 2020; Sepasgozar, Shi et al.2020).
AI to learn and think how to solve problems, make decisions and optimise selected parameters. Some key methods are machine learning, neural networks, natural language processing, and computer vision.	While in the 1950s the concepts discussed in the literature, after the 1960s some problem-solving and symbolic methods related to computer machinery and intelligence started to emerge, such as The Imitation Game. In recent years, advances in AI have opened up new possibilities for digital twin technology.	The AI algorithms can be used to create more accurate data-driven models of physical systems.They can also be used to identify patterns and trends in large datasets, process real-time data, detect faults, diagnose system issues, and integrate multiple data sources.

the following years. Sepasgozar (2020) defined the digital twin in the city context and reviewed the evolution of the definition over the years. However, Table 1.2 also shows a summary of the shift in digital twin definitions that can provide functionality.

1.3 Big data and digital twin components

A typical digital twin comprises several key components, such as sensing tools to acquire data, followed by mechanisms capable of computing Big Data generated from physical systems' movements and changes. The results of this computation can then be visualised or represented in a virtual environment. Visualisation, modelling, and dashboard systems play a crucial role in converting computed data

Table 1.2 Comparison of basic and advanced versions of digital twins for different purposes

Initial versions	*Recent versions*	*Note*
Static models	Dynamic models (Barthelmey, Lee et al.2019)	Static digital twins are created based on a snapshot of a physical system (Enders 2022) or a tangible aspect of an object at a specific point in time.Dynamic twins can be updated in real-time with data from sensors and other sources representing full aspects of the physical system.
Single components systems	Complex systems such as city digital twins, supply chain twins, healthcare twins and the Enterprise Digital Twin (EDT) (Kulkarni, Barat et al.2022).	Single-component digital twins may represent a single physical object or device, or an individual component of a system.Enterprise digital twins are used to represent an entire organisation or ecosystem, and a twin of a system represents an entire complex system or network of physical objects, product development, or changes in materials (Sytov, Vakhranev et al.2021).
Product or asset for single-stage use	Process or maintenance for the lifecycle (Lifecycle Twin)	Many examples represent a product, such as a machine, an asset, or a component, but the demand for representing processes and optimisations, such as a supply chain and excavation processes, is increased.Physics-based models often use fundamental laws of physics to represent the behaviour of physical objects, but data-driven models utilise historical and real-time data to model behaviour for prediction purposes (Goodwin, Xu et al.2022).
Reactive	Proactive	Reactive digital twins are used to represent the physical entity, monitor how it works, and diagnose issues of the physical product in real-time, while proactive digital twins are used to predict (Enders 2022) how the system will work in the future, optimise the process, and prevent issues before they occur (Palchevskyi and Krestyanpol 2020; Alkhateeb, Jiang et al.2023).

into usable or actionable information. These tools are instrumental in addressing real-world challenges and aiding human decision-making, especially in scenarios where the volume and complexity of data are beyond the processing capabilities of the human brain. Figure 1.2 shows various common components of digital twins.

Digital twins will become more important technologies in managing numerous incidents and complexities in urban or other intricate environments. For example,

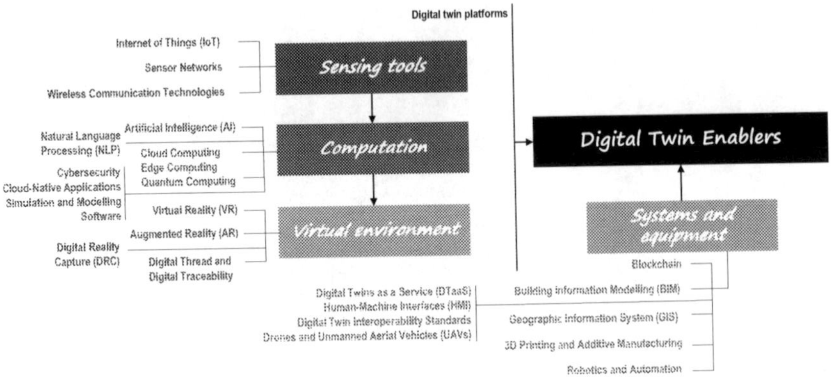

Figure 1.2 Enabling tools and systems associated with digital twins.

the integration of the Internet of Things (IoT) and other enablers may address sustainability, including optimising energy, enhancing the resilience of physical systems and smart cities, helping organisations to reduce their environmental impacts, and increasing the adoption of digital twins in some sectors, such as manufacturing and smart cities. There is also the emergence of platforms to support digital twinning, develop interoperability and standards for it, and the advancement and utilisation of AI and machine learning. These activities can be addressed by various twins, such as product or asset twins, process or system twins, and performance or experience twins.

Figure 1.2 also refers to some common systems or tools used in the smart city, additive manufacturing, or construction contexts, such as:

- Blockchain: useful, since it offers a secure digital twin with the capability of traceability, compliance, authenticity, privacy, permissions, and decentralisation (Hasan, Salah et al. 2020; Yaqoob, Salah et al. 2020; Teisserenc and Sepasgozar 2021).
- Building information modeling (BIM): useful to create and manage digital elements of buildings and infrastructure assets, enabling digital twins to make the outcome usable to construction practitioners for achieving better design, construction, and maintenance (Honghong, Gang et al. 2023; Sepasgozar, Khan et al. 2023).
- Geographic information system (GIS): useful to collect and analyse geospatial and temporal data about the city environment, such as land use, topography, and utility locations, and to optimise city planning and site preparation (Visner, Shirowzhan et al. 2021; Shi, Pan et al. 2023).
- 3D Printing and additive manufacturing: used mainly in building construction to create walls and structures, such as architectural models and building components, faster and more cost-effectively. They can be integrated with digital twins for automated operation and smart additive manufacturing (Rachmawati, Putra et al. 2023).

- Robotics and automation: can be used to carry out repetitive and risky tasks, such as material handling and excavation; the integration of industrial robotics, avatars, or multiple robotic systems with digital twins may offer greater accuracy, productivity, and efficiency (Fait and Mašek 2023; Huang, Tlili et al. 2023; Niermann, Doernbach et al. 2023).

Table 1.3 presents some functions that are expected from a digital twin and the relevant commercial platforms that may provide these functions. Since commercial tools are advancing quickly, their websites and technical documents are the best reference to learn about their technologies or tools. However, Table 1.3 gives an overall idea of the possibilities and state-of-the-art in the context of the current standard practices.

1.4 AI and digital twin evolvement

AI significantly affects the advent of digital twin practice, and will play an important role in evolving data analytics, learning, thinking, and predicting events. Some of the potential advancements of digital twins are listed as follows:

- **Efficient data analytics**: Digital twins can be efficient if vast amounts of data are available to establish accurate simulations and digital models of physical environments, including numerous objects. AI offers capabilities for Big Data analytics by automating the digital data collection process through sensing technologies and analysing the data in real time.
- **Real-time performance optimisation and prediction**: Digital twins can optimise physical assets' performance by predicting how they will perform under different conditions. AI-based analytics can recognise patterns and trends of changes in physical assets over time, which may be difficult for human users to detect. This can greatly enhance the data analysis capabilities of a digital twin application, offering more accurate and comprehensive models of physical systems. AI can help improve prediction accuracy by analysing data in real time and adjusting models accordingly.
- **Automated maintenance**: Digital twins can analyse the events, predict incidents and give warnings when the assets in operation need maintenance or repairs. However, AI enhances the automatic process of predicting the maintenance time and suggests when and how the repair should be carried out by using advanced machine learning algorithms to analyse detailed data, and identify patterns that may indicate a need for maintenance.
- **Enhancing simulation capabilities**: The availability of big data can help simulate physical systems' behaviour under different conditions. For example, the behaviour of an asset under various temperatures and thermal conditions or the behaviour of a vehicle at various speeds, road surface materials, terrain, and weather conditions. AI may enhance the accuracy of these simulations by creating more detailed models and identifying areas where improvements can be made. The simulation can consider the behaviour of buildings and

Table 1.3 Functionality and capability of digital twins using commercial platforms

No	*Expected functions of standard tools for creating and enhancing digital twins*	*Examples of relevant commercial platforms*[a]
1	Model buildings, frames, energy networks, and infrastructure all monitor, analyse, and predict the performance of connected environments, optimise assets and systems, and support integration with IoT devices.	Microsoft Azure Digital Twins
2	Create and manage digital twins, optimise their performance and maintenance, simulate, and validate asset properties, support real-time data analytics and machine learvvning, and support how twins think and perform autonomously using AI.	Siemens Digital Twin
3	Simulate complex systems, including mechanical, electrical, and fluidic components.They have enhanced fusion capabilities, thermal power libraries, optimise system performance, predict failures, and support integrating physics-based and data-driven models.	ANSYS Twin Builder
4	Manage digital twins of infrastructure assets such as bridges, dams and roads and make them sustainable and resilient, improve performance, and reduce maintenance costs using AI.They offer a single view of truth systems, facilitate the use of large-scale datasets, and support collaboration across multiple stakeholders and disciplines.	Bentley iTwin Platform
5	Connect people, ideas, data, and the processes of a business or organisation and their stakeholders and support real-time collaboration to transform initial designs into innovative assets or solutions.	Dassault Systèmes 3DEXPERIENCE
6	Manage IoT and connect devices, sensors, gateways, and systems, giving access to real-time data analytics on a cloud-hosted platform with secure data communication for reliable predictive maintenance and anomaly detection.	IBM Watson IoT platform
7	Manage complex systems and equipment and perform what-if analysis. It supports real-time monitoring and integration of applications by using an Integration Cloud Service (ICS), enabling prediction of maintenance and simulation-based optimisation.	Oracle IoT digital twin
8	Monitoring and analysis of the performance and health of assets based on benchmarks or predefined thresholds, predict maintenance, and help improve reliability and track production events.	GE digital asset performance management (APM)

[a] *Note:* Refer to relevant commercial websites for updated information.

infrastructure, allowing designers to study how different design alternatives and materials may affect energy efficiency, durability, and safety and consequently, how it may improve the performance of the physical asset, whether a building or infrastructure.

- **Enabling autonomous thinking and decision-making**: Incorporating AI into digital twin applications can enhance the efficiency and capability of independent mechanisms for learning, thinking, making decisions, and taking appropriate actions that reduce costs and improve sustainability, productivity, and safety. For example, AI can create digital twins of heating, ventilation, and air conditioning (HVAC) systems (Fathollahi and Andresen 2023) in a building to monitor and optimise their performance in terms of hours, months, and seasons. AI will eventually enable the digital twin to make autonomous decisions and take actions based on real-time sensor data. For example, the digital twin may adjust the temperature or humidity levels based on occupancy, daylight, or weather conditions.

In summary, this chapter provides an overview of digital twin concepts and highlights some of the previous practices and relevant applications. As the technological ecosystem evolves and becomes more accessible to various industries, the primary focus of innovators should shift towards developing practical applications and use cases. The industry expects to have access to those technologies that offer reliable solutions to current challenges.

References

Alkhateeb, A., S. Jiang and G. Charan (2023). "Real-time digital twins: Vision and research directions for 6G and beyond." *IEEE Communications Magazine*.

Barthelmey, A., E. Lee, R. Hana and J. Deuse (2019). *Dynamic digital twin for predictive maintenance in flexible production systems*. IECON 2019 – 45th Annual Conference of the IEEE Industrial Electronics Society, IEEE.

Enders, M. R. (2022). *Understanding and applying digital twins-results of selected studies*, Friedrich-Alexander-Universitaet Erlangen-Nuernberg.

Fait, D. and V. Mašek (2023). Digital twins for industrial robotics: A comparative study. *Design, simulation, manufacturing: The innovation exchange*. Vitalii Ivanov, Justyna Trojanowska, Ivan Pavlenko, Erwin Rauch, Ján Piteľ, Springer: 26–35.

Fathollahi, A. and B. Andresen (2023). "Multi-machine power system transient stability enhancement utilizing a fractional order-based nonlinear stabilizer." *Fractal and Fractional* **7**(11): 808.

Goodwin, T., J. Xu, N. Celik and C.-H. Chen (2022). "Real-time digital twin-based optimization with predictive simulation learning." *Journal of Simulation* **18**(1): 1–18. DOI: 10.1080/17477778.2022.2046520

Hasan, H. R., K. Salah, R. Jayaraman, M. Omar, I. Yaqoob, S. Pesic, T. Taylor and D. Boscovic (2020). "A blockchain-based approach for the creation of digital twins." *IEEE Access* **8**: 34113–34126.

Honghong, S., Y. Gang, L. Haijiang, Z. Tian and J. Annan (2023). "Digital twin enhanced BIM to shape full life cycle digital transformation for bridge engineering." *Automation in Construction* **147**: 104736.

Huang, R., A. Tlili, L. Xu, C. Ying, L. Zheng, A. H. S. Metwally, D. Ting, T. Chang, H. Wang and J. Mason (2023). "Educational futures of intelligent synergies between humans, digital twins, avatars, and robots-the iSTAR framework." *Journal of Applied Learning & Teaching* **6**(2): 1–16.

Kerin, M. and D. T. Pham (2019). "A review of emerging Industry 4.0 technologies in remanufacturing." *Journal of Cleaner Production* **237**: 117805.

Kulkarni, V., S. Barat, T. Clark and B. S. Barn (2022). *Digital twin as an aid for decision-making in the face of uncertainty*. 2022 Winter Simulation Conference (WSC), IEEE.

Niermann, D., T. Doernbach, C. Petzoldt, M. Isken and M. Freitag (2023). "Software framework concept with visual programming and digital twin for intuitive process creation with multiple robotic systems." *Robotics and Computer-Integrated Manufacturing* **82**: 102536.

Palchevskyi, B. and L. Krestyanpol (2020). *The use of the "digital twin" concept for proactive diagnosis of technological packaging systems*. International Conference on Data Stream Mining and Processing, Springer.

Rachmawati, S. M., M. A. P. Putra, J. M. Lee and D. S. Kim (2023). "Digital twin-enabled 3D printer fault detection for smart additive manufacturing." *Engineering Applications of Artificial Intelligence* **124**: 106430.

Saracco, R. (2021). "Digital twins' future." Retrieved 25/11/2023, 2023, from https://cmte.ieee.org/futuredirections/2021/01/26/digital-twins-future/.

Sepasgozar, S. M. (2020). "Digital technology utilisation decisions for facilitating the implementation of Industry 4.0 technologies." *Construction Innovation* **21**(3): 476–489.

Sepasgozar, S. M. E. (2020). Digital twin and cities. In *The Palgrave encyclopedia of urban and regional futures*. Robert Brears (Editor-in-Chief), Springer International Publishing: 1–6.

Sepasgozar, S. M., A. Shi, L. Yang, S. Shirowzhan and D. J. Edwards (2020). "Additive manufacturing applications for Industry 4.0: A systematic critical review." *Buildings* **10**(12): 231.

Sepasgozar, S. M., A. A. Khan, K. Smith, J. G. Romero, X. Shen, S. Shirowzhan, H. Li and F. Tahmasebinia (2023). "BIM and digital twin for developing convergence technologies as future of digital construction." *Buildings* **13**(2): 441.

Shi, J., Z. Pan, L. Jiang and X. Zhai (2023). "An ontology-based methodology to establish city information model of digital twin city by merging BIM, GIS and IoT." *Advanced Engineering Informatics* **57**: 102114.

Smith, K. and S. Sepasgozar (2022). "Governance, standards and regulation: What construction and mining need to commit to Industry 4.0." *Buildings* **12**(7): 1064.

Sytov, A., A. Vakhranev and F. Ereshko (2021). *Enterprise digital twin research*. 2021 14th International Conference Management of Large-Scale System Development (MLSD), IEEE.

Teisserenc, B. and S. Sepasgozar (2021). "Project data categorization, adoption factors, and non-functional requirements for blockchain based digital twins in the Construction Industry 4.0." *Buildings* **11**(12): 626.

Visner, M., S. Shirowzhan and C. Pettit (2021). "Spatial analysis, interactive visualisation and GIS-based dashboard for monitoring spatio-temporal changes of hotspots of bushfires over 100 years in New South Wales, Australia." *Buildings* **11**(2): 37.

Yaqoob, I., K. Salah, M. Uddin, R. Jayaraman, M. Omar and M. Imran (2020). "Blockchain for digital twins: Recent advances and future research challenges." *IEEE Network* **34**(5): 290–298.

2 Organisational factors affecting digital technology adoption in the architecture industry

A systematic literature review

Hesham Algassim, Samad M. E. Sepasgozar, Michael J. Ostwald and Steven Davis

2.1 Introduction

Despite the benefits of using advanced digital technologies (Eastman et al. 2011; Attaran and Celik 2023), the architecture, engineering, and construction (AEC) sectors have been described as both slow and reluctant to adopt them. Liu, Lu et al. (2018), for example, affirm that the adoption of technologies in architectural practices is not straightforward, and there is often considerable reluctance from industry users, a significant problem that has not yet been resolved. Past research (Maclennan 2014) has identified a range of possible factors explaining this reluctance, typically focusing on perceived ease of use, the complexity of the technology, technology cost, and customer and vendor-related attributes. While these are valuable, they ignore factors specific to the architecture industry and its organisational types. For example, it is apparent that the problems of technology adoption are not universal, and some countries have displayed a growing readiness for building information modelling (BIM) technology (Chen et al. 2015; Doumbouya 2016). Despite this, a report by Assemble (2015), an Autodesk company, argues that embracing technologies successfully is still a challenge. In that survey, it is evident that the industry has become more accepting of technology in recent years, with employees being slightly more willing to adopt new ones than the managers of firms. Specifically, 27% of employees and 32% of managers are hesitant about adopting new technology (Assemble 2015), a reluctance attributed to the high cost of information technology (IT). Margins in construction projects are also notoriously thin, which makes purchasing and deploying new software relatively difficult to achieve (Ahuja et al. 2020). The gap between employees' and managers' attitudes to technology identified by Assemble is reflected in other studies (Lines 2017), which argue that bridging the gap between the "office" (organisation management) and the "field" (actual end-users of technology) is increasingly important. Firms with a capacity to manage technological change more effectively

DOI: 10.1201/9781003507000-2

can position themselves as early adopters and may be able to reap the benefits of using the technology. For this reason, technology adoption is often directly tied to the productivity, efficiency, and economic performance of architectural firms.

This brief background identifies several knowledge gaps that are the catalyst for the present research. First, there are currently no rigorous literature reviews about technology adoption, specifically in the architectural context. Second, there is a need to explore and establish the relationships between the attitudes of end-users and managers to determine if these are contributing factors to the slow adoption of technology in architecture. These can be rephrased as two research questions about architectural practices:

i What are the organisational factors influencing the process of technology adoption?
ii Does the relationship between the end user, customer and managerial executives affect the managerial decision-making process to adopt digital technology?

The further gap is about the capacity of new knowledge developed from answering these questions to improve the architecture industry's capacity to adopt digital technologies.

In response to these knowledge gaps, the present chapter uses a systematic literature review to answer these questions and then develop a new integrated Technology Adoption Model specifically for the architectural service sector. As such, this chapter proposes one primary theoretical contribution through the application of a systematic literature review, which is useful for the exploration of organisational factors affecting technology adoption within architectural firms, focused on the twin issues of "perceived ease of use" and "perceived usefulness."

2.2 Methods

By rigorously collecting and reviewing past knowledge developed through related but different studies, a researcher can gain a comprehensive overview of knowledge development in a particular field. The systematic literature review is a rigorous, auditable, and trustworthy methodology that collects and combines existing knowledge to develop new knowledge (Irani and Kamal 2014). The method is valuable because it can assist a researcher in recognising gaps in past research and suggest recommendations and opportunities for future research. Focusing such a literature review on well-structured taxonomies related to the subject of interest allows for the meaningful clustering of knowledge. In addition, the systematic literature review provides a means of expanding and generalising existing knowledge (Irani and Kamal 2014). The systematic review for this chapter was conducted in 4 stages (see Figure 2.1).

2.2.1 Stage 1: Planning stage

The research questions that provide the focus of the systematic review respond to the past literature that identifies the importance of organisational and role-specific factors inhibiting technology adoption (Assemble 2015; Lines 2017; Ahuja et al.

2020). The first research question asks what the organisational factors influencing the process of technology adoption are. The second question asks, does the relationship between the end user, customer and managerial executives affect managerial decision-making about technology adoption?

2.2.2 *Stage 2: Execution stage*

The two research questions and the initial reports, which were the catalysts for them, provided the keywords for the database search of titles and abstracts in Science Direct, Scopus, Engineering Village, and Web of Science. Keyword combinations came from "Digital technology AND Technology Adoption AND Architecture," "Digital innovation AND Perceived Usefulness," "Technology Acceptance AND Digital Innovation AND AEC industry," "Technology Diffusion AND Digital Access AND Literature Gaps," "Industry 4.0 AND Architecture Industry," and "Digital technology Adoption AND Organisational Factors." First, the keywords for searching should combine at least one term in each category. Second, a 13-year period (2005–2020) was chosen to deliver recent results because architecture is an evolving industry.

2.2.3 *Stage 3: Review results*

All identified papers ($n = 100$) were screened using inclusion and exclusion criteria, additional sources cited in these papers were also considered ($n = 5$), and duplicates ($n = 10$) were removed. The included papers were passed through a quality check based on their title and content to ensure that they contributed to the specific research base ($n = 70$). In the extraction stage, the papers are synthesised for findings and evidence that relate to each of the research questions ($n = 53$).

2.2.4 *Stage 4: Data review*

The 53 papers remaining at the end of the systematic literature review all offer information that directly contributes to answering the two research questions. The papers reinforce the conceptual modelling by offering technological, cost-related, and culture-related information that affect the creation of affordable and sustainable architectural models. The data was later used as a basis for the development of the digitised architectural design methodology.

2.3 Results

The majority of the initial 100 (pre-filtering) papers identified through the review were published in the years 2018, 2019, 2020, and 2022, respectively (Figure 2.2). From the figure, it is apparent that most papers came from the year 2022.

A review of the 53 papers reveals both recurring themes and repeated gaps in knowledge about the acceptance and adoption of digital technology in the AEC sectors. One dimension of past research, which highlights both a theme and gap, is the extent to which all three sectors are treated as being identical, despite evidence

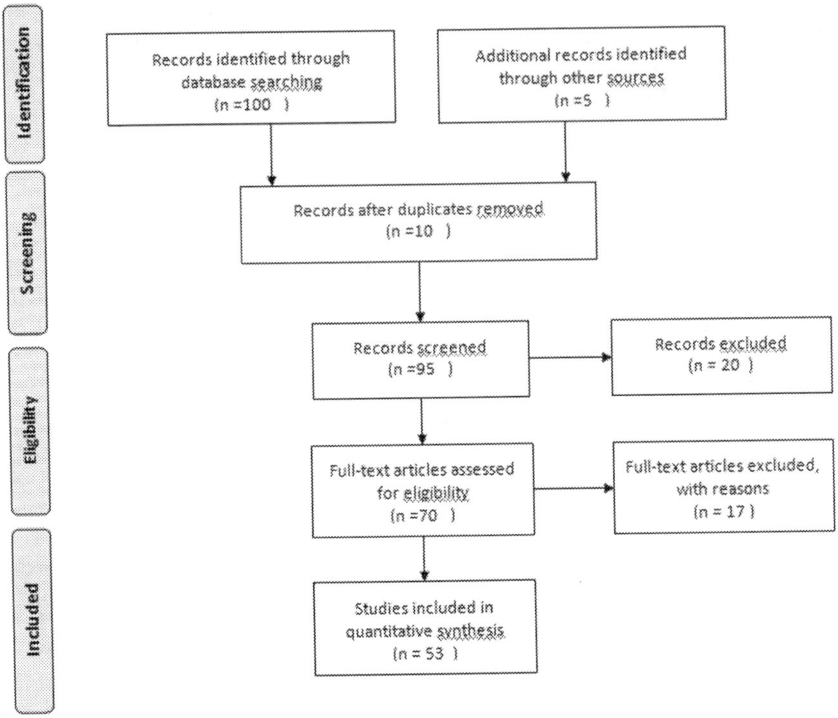

Figure 2.1 Systematic review protocol.

to the contrary. For example, most of the research (Bin Zakaria et al. 2013; Chen et al. 2015; Eadie 2015; Rogers 2015; Doumbouya 2016; Hong et al. 2016; Sepasgozar 2016; Alumayn 2017; Almuntaser 2018; Li et al. 2019; Terreno 2019; Ahuja et al. 2020; Mehrbod 2020) is focused on the barriers to technology adoption across the entire AEC industry, resulting in a generalisation that the factors causing a lag in technology adoption are similar across all three. This is potentially problematic as each of these industries has unique features. Further, the focus of technology adoption on BIM alone has (Bin Zakaria et al. 2013; Chen et al. 2015; Eadie 2015; Rogers 2015; Doumbouya 2016; Hong et al. 2016; Sepasgozar 2016; Alumayn 2017; Almuntaser 2018; Li et al. 2019; Terreno 2019; Ahuja et al. 2020; Mehrbod 2020) limited the literature on other digital technologies used in the whole life architecture project cycle.

A review of the 53 papers also reveals almost no research into the topic of technology adoption to support brief preparation, which is an important stage of the architecture project cycle. According to Hudson (2010), brief preparation is a step that establishes an understanding of the project needs between both the clients and architects. An understanding of how digital technology can be effective in supporting the process of brief preparation is lacking in the literature. According to Chen et al. (2015) digital technologies related to Big Data and IoT can help streamline

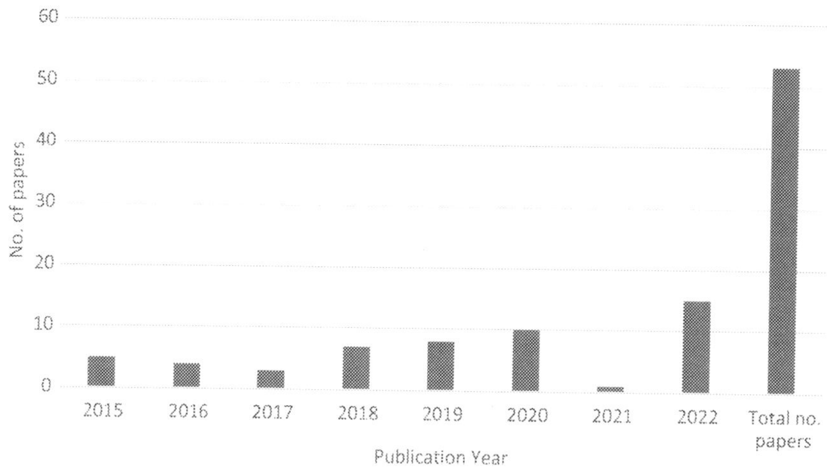

Figure 2.2 Distribution of papers across years.

data management by creating relevant output during the initial planning and design stages in the architecture industry. Existing information modelling software can help bridge the gap between the project investors' and architects' understanding of project needs and requirements, reducing conflicts in the early design stages (Bin Zakaria et al. 2013; Chen et al. 2015; Eadie 2015; Rogers 2015; Doumbouya 2016; Hong et al. 2016; Sepasgozar 2016; Alumayn 2017; Almuntaser 2018; Li et al. 2019; Terreno 2019; Ahuja et al. 2020; Mehrbod 2020). Brief preparation is, however, rarely mentioned in the past literature, and there appears to be a lack of understanding of its significance (Eastman et al. 2011).

The role of architectural managers, who are the major decision-makers in technology adoption in architectural practices, is scarcely explored in the literature (Eastman et al. 2011). Several of the past papers do, nevertheless, identify TAM as a useful model for understanding socio-technical aspects of adoption (Bangdome-Dery 2013; Kassem 2015; Rogers 2015; Sepasgozar 2016). TAM, a model based on human behaviour, has, however, tended to narrow the understanding of technology adoption to end users (Ajibade 2018). Other research considers three viewpoints in TAM: the "designer," "end user" and "client" (Bangdome-Dery 2013; Kassem 2015; Rogers 2015; Sepasgozar 2016). However, according to Lines (2017), it is the managers' attitudes that determine the speed and extent of digital technology adoption in architecture. The gap between employees' (end-users) and managers' attitudes to technology is reflected in only a few studies (Lines 2017), which argue that bridging the gap between the "office" and the "field" is increasingly important. Managers with a capacity to manage technological change effectively, can position themselves as early adopters and, as such, are able to see the rewards of using the technology.

The findings of the systematic review indicate a deficiency in existing technology acceptance model (TAM) and related models due to the incapacity to accommodate the particular needs and characteristics of the architecture industry. For example, the Green IT Adoption Model (GITAM) (Molla 2008) and SERV-QUAL (Service Quality Model) (Sunindijo 2014) models each deal with single factors, such as environmental issues and service quality, respectively.

From the review of the 53 papers, eight organisational factors affecting technology adoption are identified (Table 2.1), although none are specifically articulated in terms of the architecture industry. In the next section, these eight factors are described in more detail, with reference to the claims in the literature.

2.3.1 *Organisational factors impeding technology adoption in the architecture industry*

According to Hudson (2010), the needs of end users (the employees of a practice who use the technology) are the most important ones in architecture. Brief preparation is regarded as being among the most critical initial stages of any project. Briefs are often prepared to present a project's requirements to clients and local authorities. Hudson argues that in the process of brief development, the way a project is presented to the clients determines if the client will accept the design. The tools used to draw schematic diagrams, document detailed design stages, and designate functional performance are needed to prepare briefs and provide result demonstrability. Hudson (2010) further affirms that the architect communicates his or her intentions through the information contained in construction documents. Techlink (2010) defines brief preparation as a "dynamic process that reflects the complex interactions within the ongoing technological practice" (p. 18). A brief includes the specifications that define the requirements of a technological outcome in terms of appearance and performance. The briefs explain what needs to be done and why. The concept of result demonstrability is also tied to brief preparation in this context. Venkatesh (2003), for example, argues that individuals are more likely to perceive a technology to be useful if they can anticipate a correlation between usage and positive results. If digital technology has the potential to produce job-relevant results desired by the user, then it is more likely to be adopted.

According to Khalid (2019), in the construction industry, architects limit the amount of time they spend undertaking these tasks to ensure financial viability due to the fixed nature of project fees. As such, it becomes essential for architects to estimate the time involved in completing a project and understand their management processes and costs before commencing (Khalid 2019). The Atlanta Project Management Institute (APMI 2013) defines project time management as the processes required to manage the timely completion of the project. The belief is that it is impossible to meet a client's needs while also satisfying the myriad of project requirements with poor time management practices. According to Wang et al. (2020), the lifecycle of architecture and construction projects is usually long, and the likelihood of unexpected events occurring is also very high. Poor time management often causes misalignment between a project's actual completion time and the projected completion date.

The influence of environmental factors on technology adoption is an area that needs further research, despite a growing body of work on this topic (Molla 2008; Akadiri 2012; Chen et al. 2015; Wang et al. 2019). Such studies, which discuss the

Table 2.1 List of eight organisational factors

Item	*Factor*	*Definitions*	*Publications examples*
1	Brief preparation	Brief preparation refers to the process of developing a guiding document that communicates to the client the intent of the project.	Hudson (2010) and Techlink (2010)
2	Result demonstrability	Result demonstrability refers to evidence-based communication capacity in brief development and presentations that affect the organisation's decision to adopt a technology.	Weidman et al.(2011) and Maruping et al.(2017)
3	Project time	Project time management is the process required to manage the timely completion of the project.	Khalid (2019), Riazi (2020), and Wang et al.(2020)
4	Environmental considerations	Environmental considerations include all aspects of the environment that impact architectural decisions in the design process.	Molla (2008), Akadiri (2012), Chen et al.(2015), Wang et al.(2019), and Tsamardinos et al.(2020)
5	Service quality	Service quality encompasses the reliability or ability to provide prompt services dependably and accurately.	Bitner et al.(2010), Maclennan (2014), Sunindijo et al. (2014), Rogers (2015), Aleksandrova et al.(2019), and Büyüközkan et al. (2020)
6	Cost-effectiveness	Cost-effectiveness refers to the ability to effectively manage, control, and estimate the cost in the process, allowing the organisation to maximise the digital technology's functionality.	Hojjati (2016), Nnaji (2018), and Mehrbod (2020)
7	Training considerations	Training considerations refer to any factor related to training that may affect the process of technology adoption, such as additional training time and other costs of training.	Ghobakhloo et al.(2011), Bin Zakaria et al.(2013), and Li et al.(2019)
8	User-friendliness	User-friendliness refers to the ease with which a user can operate a technology to achieve the desired outcome.	Ghobakhloo et al.(2011) and Kim et al.(2013)

effect of environmental factors on technology adoption as features (Tornatzky 1990; Molla 2008), were instrumental in the formulation of GITAM, which posits that the environmental (industry and regulatory) contexts of an organisation can either encourage or inhibit the adoption of a given technology. Wang et al. (2019) supports

Molla (2008) contention that the study of the environmental influence on technology adoption is essential. According to Wang et al. (2019), the increased requirements around sustainability have made architectural stakeholders more aware of the need to consider the greening ability of technology. Wang et al. (2019), affirm that most organisations and architects are adopting green building technologies to prevent environmental problems caused by construction failures. Wang et al. (2019) are unable, however, to prove their claim that environmental factors act as a driver for the adoption of greener technologies in the built environment, presenting an area of improvement in future research. Akadiri et al. (2012) note that the building industry, by virtue of its size, contributes to a high level of environmental pollution. Tsamardinos et al. (2020) observe that there is a growing need among organisations committed to environmental sustainability for better strategies and technologies to be adopted in the building industry. According to Akadiri (2012) architects involved in the building process can facilitate environmental sustainability by adopting the right technologies during the design stage of the building project.

The importance of service effectiveness for technology adoption is identified in several studies. For example, Bitner et al. (2010) argue that customers and employees are more effective in receiving and providing services through the use of digital technology. Bitner et al. (2010), however, recognise that there are still gaps existing in service delivery. A design and standards gap is evident, for example, when organisations have not managed to deliver design services that meet actual standards (Bitner et al. 2010). Sunindijo et al. (2014), consider service quality to be an essential factor that affects behavioral intention and client satisfaction. Organisations in the construction sector need to continually improve their service quality to retain their clients and ensure their survival in a fast-paced industry (Büyüközkan et al. 2020).

Service quality is an often intangible characteristic in the construction industry, given that a construction project starts from an abstract design and attains tangibility and quality primarily once the construction is complete (Bitner et al. 2010). Service quality is defined as "the reliability or the ability to perform the promised service dependably, accurately and in high quality" (Sunindijo et al. 2014). The SERVQUAL model has been widely used to assess the quality of architectural service, including being employed to identify items that indicate good service quality and assess clients' perceptions of service quality provided by construction professionals (Sunindijo et al. 2014). Given the importance of service quality in the architecture industry and its organisation, there is a need to explore its impact on the technology adoption process.

According to Hojjati (2016), the results of Cost-Benefit Analysis (CBA) and Cost-Effectiveness Analysis (CEA) play a significant role in decision-making with regard to investing in technology. While various researchers have explored factors affecting technology adoption, the cost-effectiveness of technology for completing a project remains under exploration. One example, Hojjati (2016), attempted to analyse the cost-effectiveness of technology by focusing on the concept of software building. Hojjati (2016) further claimed that the use of advanced IT technology in architecture would result in overall cost savings in a given project. However,

Hojjati's (2016) study is limited to the concept of software building at large, which may have led them to under-evaluate the cost-effectiveness of technology in depth in specific architecture projects. According to Mehrbod (2020), the cost of design can be very high if not well managed. Proper top-down management of architecture projects can save on costs, allowing an organisation to realise more benefits from the use of technology.

Training considerations refer to any factor related to ease of use that may affect the process of technology adoption, such as additional learning time or costs of professional training. Some past research explores this concept, although often only in brief (Li et al. 2019). For example, a study trying to understand the adoption of BIM technology in the architecture industry in China identified the lack of training as a factor contributing to the low rate of adoption. The successful adoption of BIM technology required organisations to invest in training, and firms were reluctant to invest (2019). Bin Zakaria et al. (2013) confirm the views of Li et al. (2019) that organisations and firms were reluctant to invest. Li et al. (2019) argue that training is a vital factor in the success of implementing new technologies in building industry firms. Li et al. (2019), however, fail to include training considerations as a factor affecting technology adoption. Conversely, Ghobakhloo et al. (2011) study training within the end-user context but do not discuss how it affects managerial-level decision-making around adopting technology.

User-friendliness, the final theme, refers to the ease with which a user can operate software to achieve the desired outcome. In exploring the inadequacy of user interfaces, research (Kim et al. 2013) suggests that technology used in construction (such as CAD) held no appeal to the end-user because of poor user interfaces. According to Kim et al. (2013), the more useable the software, in terms of its functionality, the more acceptable it is to end-users. Kim et al. (2013), further note that construction industry organisations typically prioritise the deployment of software that is easy to use. This suggests the need for the industry to invest in understanding the impact of user-friendliness on technology adoption Kim et al. (2013). In studying the adoption and use of IT in a small-sized organisation, Ghobakhloo et al. (2011) identify user-friendliness as a critical attribute that influences the adoption of technology. While that study's findings are in line with those of other researchers, they are not valid for the whole construction industry, as the research focused on small-sized organisations, leaving out large and medium-sized construction firms.

2.4 Future directions

An intensive critical review of the 53 papers resulted in identifying eight factors influencing technology adoption that should be examined in the design context. Furthermore, Table 2.2 shows the outcome of the critical content review that resulted in three major themes representing important gaps in the literature on technology adoption. The gaps identified highlight the need to bridge the awareness of the existence of digital technology and its actual adoption to increase efficiency in the architecture industry. The themes, deficiencies, and future directions are presented in Table 2.2.

Table 2.2 Future research directions across four key themes

Themes	*Limitations or deficiencies*	*Future directions*
Underutilisation of digital technologies	Research does not consider the use of digital technologies, such as information modelling technologies, in the early stages of architecture projects, and especially in the process of brief preparation.	More research on how digital technologies related to big data, IoT, and Information modelling tools can be used in information management to aid processes such as brief preparation need to be performed.
Lack of recognition of the manager's perspective in technology adoption	Most research is focused on the designer, end user, and client's view, ignoring the manager's role and influence on technology adoption processes.	There is an urgent need for future research to include the manager as an important component, as the managerial decision-making process is often the final determinant of technology adoption in organisations.
Technology adoption in the architecture service sector	An architecture technology adoption framework/model that includes industry-specific factors affecting technology adoption is needed.The result is a poor understanding of the technology adoption process in the architecture industry.	Future research will need to propose a model that can be tested empirically to determine its functionality in fully understanding the process of technology adoption in the architecture industry.

2.5 Conclusion

This chapter presents the results of a systematic literature review on technology adoption, largely for the architecture service industry.

In response to the first research question asked in this chapter – *what are the organisational factors influencing the process of technology adoption?* – this chapter identifies eight factors. The key factors hindering technology uptake by architectural firms are Brief Preparation, Result Demonstrability, Project Time, Environmental Considerations, Service Quality, Cost-effectiveness, Training Considerations, and User-friendliness. Unlike past studies focusing on individual attributes and technical functionalities, architectural firms are likely influenced by specific factors, such as result demonstrability, project time, and service quality. The value of the proposed set of organisational factors identified in the present chapter is to help develop a conceptual framework as the basis of future investigations. These factors can be examined individually, or the relationship between them can be investigated in various countries and design firms. The chapter also identifies the limitations or deficiencies of the current literature across four main themes and offers a set of future research directions.

The second research question posed in this chapter asked, does the relationship between the end user, customer and managerial executives affect the managerial decision-making process to adopt digital technology? The literature confirms that there exists a reluctance at several levels to adopt the technology. The proper adoption of technology, including embracing change positively by the management in an architectural firm, allows that firm to better understand what is involved in the process. Yet, despite the results of past studies, the reluctance to adopt new technology in many architectural firms remains a challenge that is yet to be fully explored. In particular, the past studies of the construction industry are largely focused on the end-user level only. In contrast, this chapter shows that there is a need to explore how different factors at both user and organisational levels contribute to influencing the adoption of new technology within an organisation in the architecture service industry.

References

Ahuja, R., A. Sawhney, M. Jain, M. Arif and S. Rakshit (2020). "Factors influencing BIM adoption in emerging markets – The case of India." *International Journal of Construction Management* **20**(1): 65–76.

Ajibade, P. (2018). "Technology acceptance model limitations and criticisms: Exploring the practical applications and use in technology-related studies, mixed-method, and qualitative researches." *Library Philosophy and Practice* (e-journal). 1941. Pages 1–13 http://digitalcommons.unl.edu/libphilprac/1941.

Akadiri, P. O., E. A. Chinyio and P. O. Olomolaiye (2012). "Design of a sustainable building: A conceptual framework for implementing sustainability in the building sector." *Buildings* **2**(2): 126–152.

Aleksandrova, E., Vinogradova, V. and Tokunova, G. (2019). "Integration of digital technologies in the field of construction in the Russian Federation." *Engineering Management in Production and Services* **11**(3): 38–47.

Almuntaser, T., M. O. Sanni-Anibire and M. A. Hassanain (2018). "Adoption and implementation of BIM–Case study of a Saudi Arabian AEC firm." *Journal of Managing Projects in Business* **11**(3): 608–624.

Alumayn, S. C. E. and I. Ndekugri (2017). "The barriers and strategies of implementing BIM in Saudi Arabia." *WIT Transactions on the Built Environment* **169**(169): 55–67.

Assemble (2015). Condition and Connect BIM Data. https://construction.autodesk.com.au/products/assemble/ (Access date: 6/12/2023).

Atlanta Project Management Institute (2013). *A guide to the project management body of knowledge*, Atlanta Project Management Institute.

Attaran, M. and B. G. Celik (2023). "Digital twin: Benefits, use cases, challenges, and opportunities." *Decision Analytics Journal*, **6**: 100165.

Bangdome-Dery, A. A. K.-S. V. (2013). "Analysis of barriers (factors) affecting architects in the use of sustainable strategies in building design in Ghana." *Research Journal in Engineering and Applied Sciences* **2**(6): 418–426.

Bin Zakaria, Z., N. Mohamed Ali, A. Tarmizi Haron, A. J. Marshall-Ponting and Z. Abd Hamid (2013). "Exploring the adoption of Building Information Modelling (BIM) in the Malaysian construction industry: A qualitative approach." *International Journal of Research in Engineering and Technology* **2**(8): 384–395.

Bitner, M. J., V. A. Zeithaml and D. D. Gremler (2010). Technology's impact on the gaps model of service quality. *Handbook of service science*. Paul P. Maglio, Cheryl A. Kieliszewski, James C. Spohrer. Springer: 197–218.

Büyüközkan, G., C. A. Havle, O. Feyzioğlu and F. Göçer (2020). "A combined group decision making based IFCM and SERVQUAL approach for strategic analysis of airline service quality." *Journal of Intelligent & Fuzzy System* **38**(1): 859–872.

Chen, K., W. Lu, Y. Peng, S. Rowlinson and G. Q. Huang (2015). "Bridging BIM and building: From a literature review to an integrated conceptual framework." *International Journal of Project Management* **33**(6): 1405–1416.

Doumbouya, L., G. Gao and C. Guan (2016). "Adoption of the Building Information Modeling (BIM) for construction project effectiveness: The review of BIM benefits." *American Journal of Civil Engineering and Architecture* **4**(3): 74–79.

Eadie, R., T. Mclernon and A. Patton (2015). *An investigation into the legal issues relating to building information modelling (BIM)*. In *RICS COBRA AUBEA 2015*. Royal Institution of Chartered Surveyors. ISBN:9781783210718.

Eastman, C. M., C. Eastman, P. Teicholz, R. Sacks and K. Liston (2011). *BIM handbook: A guide to building information modeling for owners, managers, designers, engineers and contractors*, John Wiley & Sons.

Ghobakhloo, M., M. S. Sabouri, T. S. Hong and N. Zulkifli (2011). "Information technology adoption in small and medium-sized enterprises; An appraisal of two decades literature." *Interdisciplinary Journal of Research in Business* **1**(7): 53–80.

Hojjati, S. N. a. K. M. (2016). "Evaluation of factors affecting the adoption of smart buildings using the technology acceptance model." *International Journal of Advanced Networking and Applications* **7**(6): 2936.

Hong, Y., S. M. Sepasgozar, A. F. F. Ahmadian and A. Akbarnezhad (2016). *Factors influencing BIM adoption in small and medium sized construction organizations*. ISARC. Proceedings of the International Symposium on Automation and Robotics in Construction, Vilnius Gediminas Technical University, Department of Construction Economics & Property (Vol. 33, p. 1). IAARC: The International Association for Automation and Robotics in Construction.

Hudson, J. (2010). *Interior architecture: From brief to build*. Laurance King Publishing.

Irani, Z. and M. M. J. E. S. w. A. Kamal (2014). "Intelligent systems research in the construction industry." *Expert Systems with Applications* **41**(4): 934–950.

Kassem, M., B. Succar and N. B. Dawood (2015). Building information modeling: Analyzing noteworthy publications of eight countries using a knowledge content taxonomy. *Building information modeling: Applications and practices*. R. A. Issa and S. Olbina, American Society of Civil Engineers, Reston: 329–371.

Khalid, F. J. I. (2019). "The impact of poor planning and management on the duration of construction projects: A review." *Multi-Knowledge Electronic Comprehensive Journal for Education and Science Publications* **2**: 161–181

Kim, C., T. Park, H. Lim and H. Kim (2013). "On-site construction management using mobile computing technology 415–423." *Automation in Construction* **35**: 415–423.

Li, P., S. Zheng, H. Si and K. Xu (2019). "Critical challenges for BIM adoption in small and medium-sized enterprises: Evidence from China." *Advances in Civil Engineering* **2019**: 1–14.

Lines, B. C. a. R. V. P. K. (2017). "Drivers of organizational change within the AEC industry: Linking change management practices with successful change adoption." *Journal of Management in Engineering* **33**(6): 04017031.

Liu, D., W. Lu, Y. J. J. o. C. E. Niu and Management (2018). "Extended technology-acceptance model to make smart construction systems successful." *Construction Engineering and Management* **144**(6): 04018035.

Maclennan, E. A. V. B. J. P. (2014). "Factors affecting the organizational adoption of service-oriented architecture (SOA)." *Information Systems and e-Business Management* **12**(1): 71–100.

Maruping, L.M., Bala, H., Venkatesh, V. and Brown, S.A., 2017. Going beyond intention: Integrating behavioral expectation into the unified theory of acceptance and use of technology. *Journal of the Association for Information Science and Technology* **68**(3): 623–637.

Mehrbod, S., S. Staub-French and M. Tory (2020). "BIM-based building design coordination: processes, bottlenecks, and considerations." *Canadian Journal of Civil Engineering* **47**(1): 25–36.

Molla, A. (2008). *GITAM: A model for the adoption of Green IT*. ACIS 2008 Proceedings (p. 64). http://aisel.aisnet.org/acis2008/64

Nnaji, C., J. A. Gambatese and C. Eseonu (2018). "Theoretical framework for improving the adoption of safety technology in the construction industry." In *Construction Research Congress 2018*. Chao Wang, Christofer Harper, Yongcheol Lee, Rebecca Harris, Charles Berryman (pp. 356–366). ASCE library.

Riazi, S. R. M., M. N. M. Nawi and M. F. A. Yaziz (2020). "Developing a holistic project time management framework utilizing fundamental Supply Chain Management (SCM) tools to overcome delay in Malaysian public sector building projects." *International Journal of Sustainable Construction Engineering and Technology* **11**(1): 31–41.

Rogers, J., H. Y. Chong and C. Preece (2015). "Adoption of building information modelling technology (BIM) perspectives from Malaysian engineering consulting services firms." *Engineering, Construction and Architectural Management* **22**(4): 424–445.

Sepasgozar, H. Y. S. M., A. F. F. Ahmadian and A. Akbarnezhad (2016). *Factors influencing BIM adoption in small and medium sized construction organizations*. ISARC. Proceedings of the International Symposium on Automation and Robotics in Construction (Vol. 33, p. 1). IAARC: The International Association for Automation and Robotics in Construction.

Sunindijo, R. Y., B. H. Hadikusumo and T. Phangchunun (2014). "Modelling service quality in the construction industry." *International Journal of Business Performance Management* **15**(3): 262–276.

Techlink (2010). The technological practice strand: Brief development, Technology Curriculum Guide.

Terreno, S., K. Edirisinghe and C. Anumba (2019). *The influence of managerial strategy on the implementation of BIM in facilities management: A case study*. 36th CIB W78 2019 Conference: Advances in ICT in Design, Construction and Management in Architecture, Engineering, Construction and Operations (AECO). (Newcastle, United Kingdom, 18/09/2019 - 20/09/2019) (Vol. 78, p. 2019).

Tornatzky, L. G. A. F. M. (1990). "The process of technological innovation." National Science Foundation (U.S.). Division of Industrial Science and Technological Innovation. Productivity Improvement Research Section.

Tsamardinos, I., G. S. Fanourgakis, E. Greasidou, E. Klontzas, K. Gkagkas and G. E. Froudakis (2020). "An automated machine learning architecture for the accelerated prediction of metal-organic frameworks performance in energy and environmental applications." *Microporous and Mesoporous Materials* **300**: 110160.

Venkatesh, V. a. D. (2003). "A theoretical extension of the technology acceptance model: Four longitudinal field studies." *Management Science* **46**(2): 186–204.

Wang, Q., Y. J. Hu, J. Hao, N. Lv, T. Y. Li and B. J. Tang (2019). "Exploring the influences of green industrial building on the energy consumption of industrial enterprises: A case study of Chinese cigarette manufactures." *Journal of Cleaner Production* **231**: 370–385.

Wang, Q., W. Zuo and Q. Li (2020). "Engineering harmony under multi-constraint objectives: The perspective of meta-analysis." *Journal of Civil Engineering and Management* **26**(2): 131–146.

Weidman JE, Young-Corbett D, Koebel CT, Fiori C, Montague EN. 2011. Prevention through design: Use of the diffusion of innovation model to predict adoption. *International Council for Research and Innovation in Building and Construction Conference*. Billy Hare and Fred Sherratt (CIB W099 Conference 2011).

3 Digital twin maturity and readiness metrics for assessing practitioners' intention to use

Model development and multi-group structural analysis

Samad M. E. Sepasgozar, Sara Shirowzhan, Marco Mura, Alberto De Marco, Michael J. Ostwald and David Edwards

3.1 Introduction

In the 1970s, the National Aeronautics and Space Administration (NASA) utilised an early example of a digital twin (DT) during the development of the Apollo 13 project (Portela, Varsakelis et al. 2021). A DT is typically defined as the virtual representation of a physical entity and its operations (Sepasgozar 2020b), using advanced technologies such as the Internet of Things (IoT) for maintaining two-way data transfer and simulation. Many researchers describe the DT as a single model or technology, yet NASA used 15 simulators for interconnecting different elements of the Apollo mission (Gaier 2007). Thus, a DT can be a set of models interacting with different aspects of the physical entity, organisational process, or object behaviour. Significantly, NASA also assessed the maturity of advanced technologies using a specific set of metrics to measure the effectiveness of technology adoption. Such was the success of NASA's pioneering work that it has subsequently been adopted in various industrial settings, ranging from advanced manufacturing to automotive engineering and aviation (Ferreira, Biesek et al. 2021; Zutin, Barbosa et al. 2022). Other, less advanced industrial sectors, such as construction, have also begun to adopt DTs, but have hitherto failed to transfer the required technology metrics to estimate a DT application's maturity prior to acquisition. The problem with this situation is that utilising emerging technologies without a maturity assessment increases the risk of technology failure and subsequent financial loss.

In combination with building information modelling (BIM) and geographical information systems (GISs), a wide range of DT applications have been proposed or introduced in the building industry (Shirowzhan, Lim et al. 2016; Visner, Shirowzhan et al. 2021). In contrast to the Apollo mission, however, BIM is designed to be shared with identified groups of stakeholders, each with different contractual objectives (Shirowzhan, Sepasgozar et al. 2020). These stakeholders are often responsible for a specific aspect of the construction process or its logistics, using DT applications that are interoperable with their current BIM or GIS

DOI: 10.1201/9781003507000-3

datasets to achieve their goals. Thus, DT maturity in a dynamic construction context differs from other industries and requires specific support mechanisms. One of the most important is the need for relevant construction industry scales to capture technology readiness level (TRL) and maturity utilisation for the DT (Rybicka, Tiwari et al. 2016).

The need for a combined set of DT readiness and maturity scales is especially evident when a review of some of the properties of the construction sector is undertaken. For example, the construction industry has been infamously antipathetic to new technologies (Edwards et al., 2017; Parn and Edwards, 2019), and the success rate of technology adoption is dramatically reduced in the absence of assessment metrics. The development of metrics is important, particularly for DT applications, because they represent the next wave of technology evolution (within the "Industry 4.0" context) in data sharing, simulation, risk assessment, and prediction of events (Aheleroff, Xu et al. 2021).

Industry 4.0 comprises a multitude of advanced technologies that coalesce to engender industrial efficiency gains, quality control improvements, and productivity performance gains on-site and via virtual management of supporting processes and procedures (Newman et al., 2020). Fierce global competition motivates construction companies to seek innovative methods to improve the productivity and quality of their projects. Recent digital technologies – such as IoT and machine learning (ML) – encourage construction companies to develop data-oriented planning and decision-making systems and tools (Sepasgozar, Khan et al. 2023). Although Industry 4.0 has gained support internationally and is often embedded in government policy and the adoption of bespoke technology systems is in a narrow sensea burdensome process in construction as there are unknown factors involved in its implementation (Sepasgozar 2020). Therefore, measuring the technology maturity level within a construction company and its effect on organisational behaviours and intentions to adopt it are important challenges. Hicks, Larsson et al. (2009), for example, argue for the significance of technology readiness assessments suggesting that in highly competitive markets, technologies are advancing, both increasing and enhancing user expectations. Thus, the availability of benchmarks and methods for assessing the readiness of new technologies is critical (Hicks, Larsson et al. 2009). Furthermore, readiness assessment of this type is already a prerequisite in more advanced industrial sectors – such as manufacturing (Uflewska, Wong et al. 2017), training and education (Astuti, Sudira et al. 2021), maritime industries (Sullivan, Arias Nava et al. 2021), and space applications (Héder 2017) – but not in construction.

This chapter presents a DTRL framework and scale, which are adapted from the TRL model used by NASA for assessing technology maturity (Conrow 2011). The research questions addressed in this chapter through the development of the DTRL are: (i) what are practitioners' perceptions of DT maturity; (ii) what is the relationship between DT maturity and users' intentions to adopt DT; (iii) how does practitioner innovation affect their intention to use technology; and (iv) how can DT application readiness be measured? Addressing these research questions contributes to the body of knowledge by exploring the relationship between practitioners' perceptions of DT maturity and their intention to use DT. The primary theoretical contribution of this study is the DTRL framework which can be used for predicting users' intention to adopt a DT application by coupling the TRL and constructs from a echnology acceptance model (TAM). The practical benefit of the study is that it provides a DTRL scale that practitioners can use to assess the maturity of a DT application before the acquisition process.

3.2 Developing the conceptual TRL-BI model

An omnipresent concern for technology developers is whether construction companies will accept and adopt new technology (Sepasgozar 2020). Conversely, customers question whether a new technology is ready to be used, which is a factor of its maturity level (Wu, Xu et al. 2017). While past research offers conceptual models to address these questions (Ishak and Doheim 2021), a knowledge gap is associated with vendor and customer concerns (Samad, Shirowzhan et al. 2021). For example, recent studies discuss the need to examine both vendors' and customers' attitudes towards technology adoption, offering frameworks that explain different steps taken by each side (Sepasgozar 2020; Samad, Shirowzhan et al. 2021). However, these studies are built upon either a TAM for measuring individuals' perceptions or diffusion theories for categorising adopters (e.g., early or late adopters), both of which largely ignore the role of technology maturity in the acceptance process.

The TRL is a metric reflecting the maturity of a technology that is being considered for use (Sauser, Verma et al. 2006). The original NASA TRL included seven levels of maturity, which was later extended to nine. Level 9 was used by NASA (from the 1990s onwards) for examining their "flight-proven" system, for mission operations and is widely considered the most reliable scale (Mankins 1995). Most recent TRLs (mainly used in civilian practice) follow the NASA nine-level system (Straub 2021). However, ongoing efforts to modify or extend the TRL scale illustrate that the original was unable to address heightened organisational expectations for technology implementation risk analysis. In particular, the original TRL scale covers the assessment of a technology that should be operated once. However, DT applications should operate over the entire project life cycle in construction. Scholars have discussed extending the original scale to encapsulate readiness or modification to suit applications with a prolonged operational performance and where the technology is compatible or interoperable with other systems (Straub 2015). Examples of technology maturity assessment using the basic form of TRL in different contexts are scant but include Pylianidis, Osinga et al. (2021), who assessed DT applications in agriculture against three general levels based on the European Union's TRL nine-level scale. Specifically, their mapping combined the first two levels of the TRL into a conceptual phase, then levels 3–6 into a prototype phase, and finally, levels 7–9 into a deployment phase. In another example, Rybicka, Tiwari et al. (2016) used the TRL to assess recycling technologies such

Table 3.1 Constructs and measures of the conceptual TRL-BI model

Construct	*Acronym*	*Measure, reflecting an individual's technology level*
Basic principles observed and reported (Mankins 1995).	TRL1	I am aware of potential Digital Twin (DT) applications that have been identified (Klar, Frishammar et al. 2016).
Technology concept and/or application formulation (Mankins 1995).	TRL2	I believe that an initial example of the DT concept has been introduced by innovators (Klar, Frishammar et al. 2016).
Analytical and experimental critical function and/or characteristic proof-of-concept (Mankins 1995).	TRL3	I believe that proof of the DT concept was achieved in a laboratory environment (Klar, Frishammar et al. 2016).
Component and/or breadboard validation in a laboratory environment (Mankins 1995).	TRL4	I believe that the use of DT in a pilot facility was determined by innovators (Klar, Frishammar et al. 2016).
Component and/or breadboard validation in the relevant environment (Mankins 1995).	TRL5	I believe that the DT concept was proven in the pilot environment, ensuring that it will work outside the laboratory (Klar, Frishammar et al. 2016).
System/subsystem model or prototype demonstration in a relevant environment (ground or space) (Mankins 1995).	TRL6	I believe that the DT concept was proven to the extent that suggests it will work at full scale (Klar, Frishammar et al. 2016).
Demonstration in a site (Mankins 1995): System prototype demonstration in the planned operational environment.	TRL7	I believe that the DT function and its compatibility with other software have been demonstrated in a construction site (Deshmukh Towery, Machek et al. 2017).
Behavioural intention.	BI	Probability, likelihood, and willingness to adopt DTs (Song and Zahedi 2005).
	BI1	Assuming I had access to DT, I would be motivated to use it.
	BI2	Given that I had access to DT, I predict that I would use it.
	BI3	I plan to use the DT within the next 12 months.

as disposal, recovery, recycling and reuse technologies. However, such examples fail to identify how TRL scales might be relevant to the users' intention to employ the DT technology. The current study proposes a conceptual model that combines behavioural intention (BI), derived from TAM, with TRL, as shown in Table 3.1. BI is the key construct of TAM, which is widely accepted in many disciplines including construction. Figure 3.1 shows how each TRL measure may positively affect users' intentions to accept the technology. For development and testing purposes, the assumption that each of the seven TR measures positively affects the user's intention to employ a DT application can be reframed as a series of seven hypotheses (H1–7).

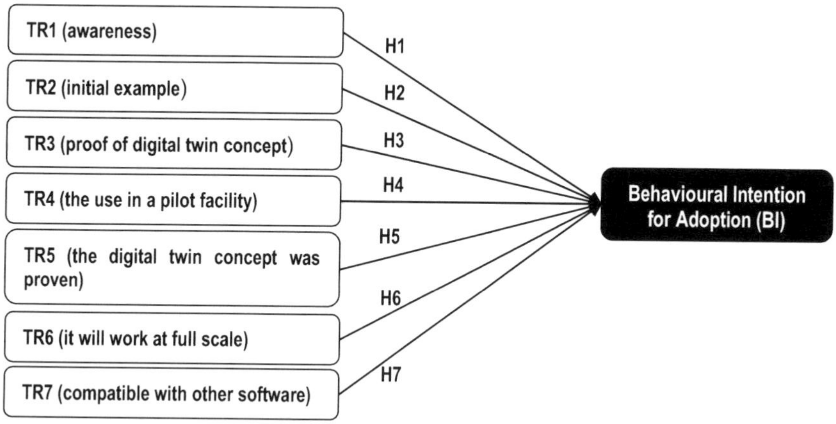

Figure 3.1 The structure of the conceptual TRL-BI model.

3.3 Research method

This research adopts a postpositivist stance and deductive reasoning (Ahmed et al., 2021; Ellis et al., 2021) to model primary data using deterministic modelling techniques (Edwards et al., 2020; Aghimien et al., 2020). A structured questionnaire was designed for collecting data related to the TRL-BI conceptual model – where the conceptual model is premised upon the literature reviewed. The data collection instrument adopted a five-point Likert scale, from 1 = strongly disagree to 5 = strongly agree, to pose >11 statements to participants. To validate this instrument, a pilot survey was conducted over a two-week period involving five experts in the field to review the relevance of the statements provided and the clarity of expression. For instance, a statement related to TRL6 (refer to Table 3.1) was changed after the pilot survey to become "I believe that digital twin concept was proven to the extent that suggests it will work at full scale."

Data was collected by distributing a survey to the target participants. Participants were firstly identified via LinkedIn by searching three different keywords within the section "People" viz: (1) DT construction industry; (2) BIM construction industry; and (3) digital technologies construction industry. Then, all profiles were double-checked to verify if they complied with the inclusion/exclusion criteria. Criteria delineated were: (1) the organisations involved could be public, private, or universities; (2) their core business must be the construction industry (architectural firm, general construction firm, facility management firm, etc.); (3) practitioners must have a managerial role in their organisations or technical expertise; and (4) practitioners must have more than three years of experience in the construction industry and implementing digital technologies (e.g., BIM). The research is limited to Australian construction companies.

Through this process, 937 profiles of practitioners were deemed eligible participants and invited to take part in the survey. The message to eligible participants included a brief description of DT and links to two videos that explained briefly what DT is. A strict ethical code of practice was adhered to during this research undertaking to secure informed consent. Specifically, participants were given: information that described the project and its objectives; assurances that the work would remain strictly confidential and all participants' data would remain anonymous; a guarantee that data would be securely stored during the research and securely disposed of after publication of results; and assurances that they could withdraw from the study at any time during the research undertaking; and an offer to access the findings when complete (cf. Fisher et al., 2018; Lau et al., 2021).

Table 3.2 Data profile and grouping participants for *t*-test analysis

Participants' attributes and various groups for comparative analysis	*Number of valid answers*	*%*
Group B: Experience		
Sub-group 1: Experience less than 11 years	74	39
Sub-group 2: Experience more than ten years	118	61
Group A: Company size		
Sub-group 1: Less than 200 employees	99	52
Sub-group 2: More than 199 employees	90	48
Group C: Innovation category		
Sub-group 1: Innovative and early adopters (categories 1 and 2, refers to Figure 3.2)	132	70
Sub-group 2: Later adopters and laggards (categories 3–5, refers to Figure 3.2)	57	30

Note: The numbers are based on completed answers to relevant questions.

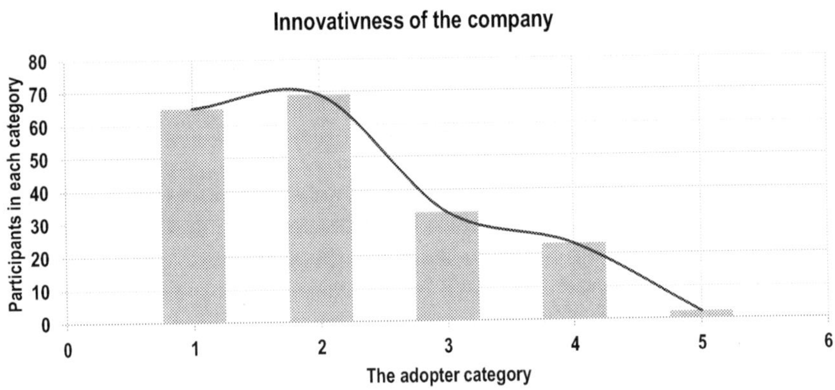

Figure 3.2 The distribution histogram of participants' categories.

Note: Categories are (1) innovative, (2) early adopter (probably 84% earlier than others), (3) a little conservative (probably adopt it earlier than 50% of others), (4) more conservative (tend to wait until 50% of similar companies have adopted), (5) resist new digital technology for as long as we can.

The survey was conducted using the Qualtrics platform. A total of 937 surveys were distributed, 234 were returned, representing a 24.97% response rate), and 189 responses were valid and used for the analysis. The data profile is shown in Table 3.2, and the distribution histogram of participants' categories in terms of being early or late adopters is shown in Figure 3.2.

3.4 Analysis and results

The partial least squares (PLS) technique was used for analysing proposed models (Ringle, Wende et al. 2015). The proposed TRL-BI model includes some relationships among factors that may affect the behavioural intention of users to use new technology. The hypothetical relationships were examined and validated by using recommended tests: "convergent validity" and "discriminant validity" (John and Reve 1982; Giancola and Zeichner 1995). Table 3.3 shows that the variance inflation factor (VIF) ranges from 1.000 to 3.026. The VIF values are <the maximum threshold of 5.0 (Henseler, Ringle et al. 2009; Hair Jr, Hult et al. 2016) – descriptive statistics, t and p values and factor loadings for BI measures are also provided. The sample means and standard deviation for other measures are 1.000 and 0.000, respectively.

Cronbach's alpha was computed to measure the internal consistency and reliability of the scale. This value is 0.817, which is well above the recommended value of 0.6 (Cronbach 1951; Hair Jr, Hult et al. 2016). The composite reliability (CR) coefficient is 0.893 for the BI, which is >the recommended value of 0.6 (Fornell and Larcker 1981). The AVE value for BI is 0.736, which reached the expected convergent validity of 0.5 (Fornell and Larcker 1981).

Table 3.4 shows the discriminant validity values for the relationship among the BIM and seven other levels of the technology readiness model. The table shows that the square roots of the AVEs are superior to the corresponding inter-variable correlations. The Fornell and Larcker (1981) criteria were used to test the data's independence and level of correlation (Sepasgozar 2022). The discriminant validity matrix's diagonal cells are displayed in Table 3.4, along with the correlation between latent variables, which exceeds the suggested cut-off point of 0.7 and ranges from 0.858 to 1.0. According to Table 3.4, every variable has an individual reliability coefficient greater than 0.4, which is higher than the minimally acceptable level that is suggested for exploratory research (Hulland 1999). All of the measurement models' tests taken together demonstrate that the model's divergence validity and the measures' convergent and discriminant validity were appropriate.

Bootstrapping analysis was conducted to validate the relationship among variables. Table 3.5 reports upon the results of statistics (including a two-tailed t-test from the bootstrapping) and reveals that the inner VIF is <5, ranging from 1.644 to 2.467, indicating that collinearity is acceptable. Table 3.5 shows statistics for seven hypotheses, with H7 strongly supported by the sample data. However, the outcome could differ for sub-groups created based on participants' attributes such as company size, experience, and innovativeness. Consequently, independent t-tests were carried out for all paths and sub-groups, and the outcomes are visualised in Figure 3.3.

Table 3.3 Descriptive analysis including t- and p-values, loadings, and collinearity statistics (VIF)

ID	*Original sample*	*Sample mean*	*Standard deviation*	t-*Values*	p-*Values*	*Outer VIF*
BI1	0.885	0.882	0.036	24.762	0.000	2.786
BI2	0.914	0.911	0.019	48.058	0.000	3.026
BI3	0.768	0.768	0.044	17.531	0.000	1.401

Table 3.4 Discriminant validity (intercorrelations) of constructs based on Fornell–Larcker criteria

	BI	*TR1*	*TR2*	*TR3*	*TR4*	*TR5*	*TR6*	*TR7*
BI	0.858							
TR1	0.282	1.000						
TR2	0.311	0.545	1.000					
TR3	0.124	0.347	0.277	1.000				
TR4	0.197	0.431	0.429	0.615	1.000			
TR5	0.334	0.336	0.417	0.530	0.597	1.000		
TR6	0.378	0.371	0.532	0.362	0.513	0.622	1.000	
TR7	0.422	0.441	0.455	0.340	0.350	0.530	0.660	1.000

Table 3.5 Summary of statistics for the hypotheses of TRL-BI, including estimated path coefficients with t-values

Path	*Original sample (O)*	*Sample mean (M)*	*Standard deviation*	t-*Values*	p-*Values*	*Inner VIF*	f^2 *effect size*
H1: TRL1 → BI	0.099	0.103	0.100	0.996	0.320	1.644	0.008
H2: TRL2 → BI	0.074	0.071	0.112	0.657	0.511	1.765	0.004
H3: TRL3 → BI	−0.108	−0.108	0.082	1.316	0.189	1.769	0.008
H4: TRL4 → BI	−0.046	−0.035	0.096	0.476	0.634	2.206	0.001
H5: TRL5 → BI	0.161	0.154	0.103	1.559	0.120	2.191	0.015
H6: TRL6 → BI	0.104	0.097	0.116	0.892	0.373	2.467	0.006
H7: TRL7 → BI	0.243	0.251	0.104	2.329	0.020	1.644	0.037

3.4.1 Comparing experienced and innovative groups

The sub-group analysis, including PLS-MGA and Welch–Satterthwaite tests, was carried out to identify the differences between sub-groups of participants or cross-verify the relationship between the technology readiness levels and the BI. The tests for each group were performed to analyse the coefficient of the seven paths of the TRL-BI by considering 5,000 permutation runs, and a two-tailed test at a significant level of $p = 0.05$. Consequently, six sub-groups were used to test the validity of the TRL-BI model (refer to Table 3.2).

An invariance test using the permutation algorithm was applied to groups A, B, and C as a prerequisite for carrying out the multi-group assessment (Henseler,

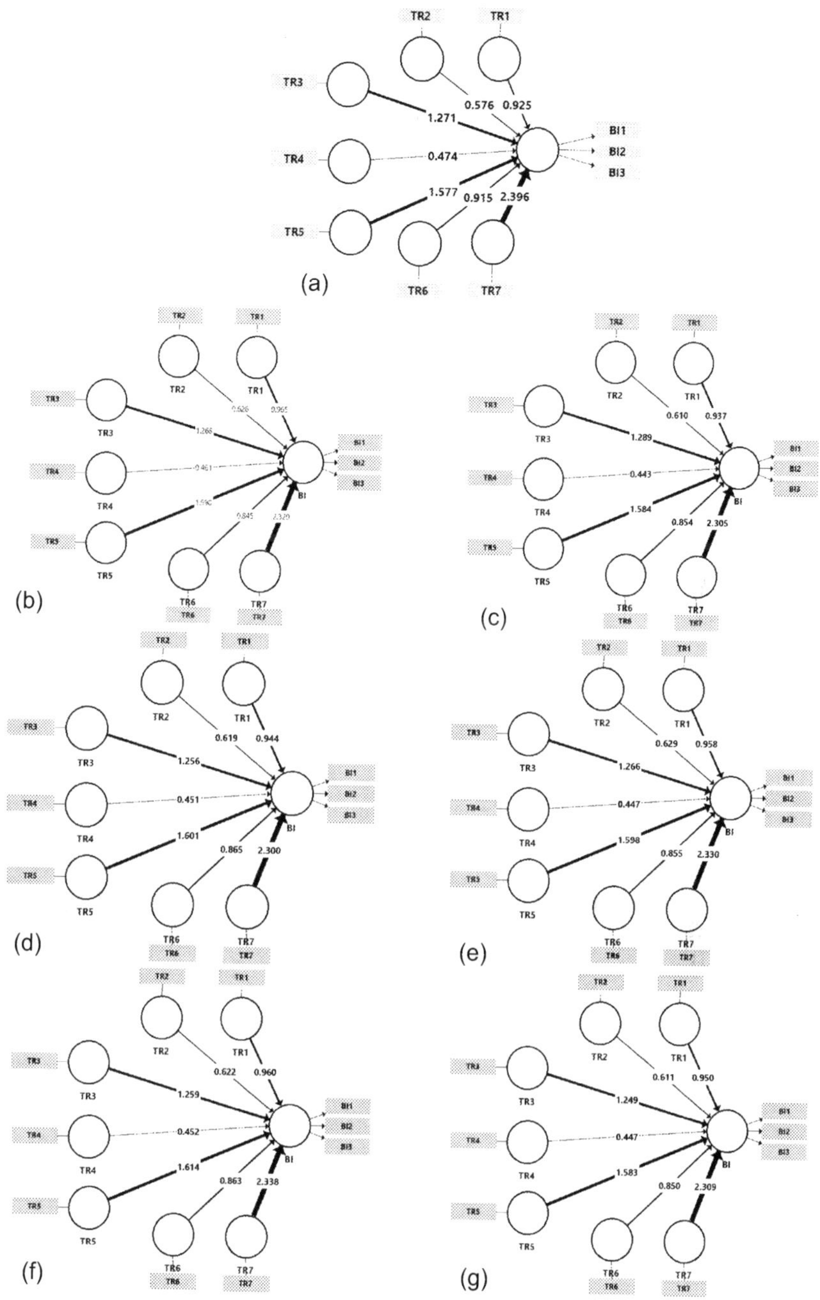

Figure 3.3 The structure of the model modifications based on different groups of participants. (a) All data set, (b) small- to medium-sized companies, (c) large companies, (d) less experienced, (e) experienced participants, (f) less innovative participants, (g) innovative participants.

Ringle Christian et al. 2016). The p-value variation for A, B, and C is 0.030–0.760, 0.036–0.086, and 0.053–0.906. In this group, the differences between H1s in sub-groups of Group A are significant. The differences between H2 and H4 in sub-groups of Group B are significant. The outcomes of every other variable in every group, aside from these fictitious routes, are non-significant (p-value > 0.05).

The PLS-MGA test results are displayed in Table 3.6, along with the p-values indicating the significant differences between Group A and B's sub-groups. The significant difference in Group A is $p = 0.029$ for H1: TRL1 to BI. In Group B, the significant differences are for H2: TRL2 to BI, H3: TRL3 to BI, and H4: TRL4 to BI, $p = 0.044$, 0.038 and 0.036, respectively. The PLS-MGA results also verify no significant differences between the rest of the groups' paths.

Table 3.7 shows the results of the Welch–Satterthwaite tests, including the t and p-values showing significant differences in the sub-groups of Group A and B. In Group A, the differences are between H1: TRL1 to BI and H2: TRL2 to BI, t = 2.226 and 2.064; and p = 0.029 and 0.042, respectively. In Group B, the significant differences are between H2: TRL2 to BI, H3: TRL3 to BI and H4: TRL4 to BI, t = 2.074, 2.123 and 2.099; and p = 0.041, 0.037 and 0.039, respectively. In Group C, the significant difference is between H6: TRL6 → BI, $t = 2.030$; and $p = 0.047$.

3.4.2 *Maturity assessment metrics development*

The literature reports various metrics suggested by scholars, each applied in a different context. Two examples of these metrics are the demand readiness level (DRL) and the manufacturing readiness level (MRL). DRL is adopted when a user or company believes that something is missing within their business. DRL refers to users' feeling that there is a gap in the market for a product (Paun 2011). It also refers to the expected functionalities of the proposed technology and responses to needed capabilities. The DRL is useful for examining market pull and users' needs. MRL is another metric to assess the maturity of a manufacturing technology, component, or system. It was designed to mitigate manufacturing risks, and assist manufacturers to identify maturity shortfalls in terms of risk identification and cost estimations. The literature argues that if a technology prototype is not mature yet, or various technology concepts are not proven, there will be limitations to what steps of the assessment level are related to the TRL steps (Ahlskog, Jackson et al. 2015).

Other than these individual metrics, some hybrid metrics can be used to calculate the maturity of a set of components that make a system. For example, the integration readiness level (IRL) is developed as a maturity measure of integrated technologies where more than two components work together, as shown in Table 3.3. The IRL is intended to be used in conjunction with TRL for estimating SRL as a measure of system maturity – refer to equation (3.1):

$$[\mathrm{SRL}]_{n \times 1} = [\mathrm{IRL}]_{n \times n} \times [\mathrm{TRL}]_{n \times 1} \quad (3.1)$$

Table 3.6 PLS-MGA results of *p*-values for all groups

Path	*Group A*	*Group B (experience)*	*Group C (innovativeness)*
H1: TRL1 → BI	0.029	0.086	0.496
H2: TRL2 → BI	0.051	0.044	0.398
H3: TRL3 → BI	0.445	0.038	0.521
H4: TRL4 → BI	0.715	0.036	0.906
H5: TRL5 → BI	0.218	0.731	0.657
H6: TRL6 → BI	0.565	0.404	0.053
H7: TRL7 → BI	0.769	0.087	0.842

Table 3.7 Welch–Satterthwaite results of *p*-values for all groups

Path	*Group A*		*Group B*		*Group C*	
	t-*Values*	p-*Values*	t-*Values*	p-*Values*	t-*Values*	p-*Values*
H1: TRL1 → BI	2.226	0.029	1.761	0.082	0.675	0.502
H2: TRL2 → BI	2.064	0.042	2.074	0.041	0.937	0.352
H3: TRL3 → BI	0.752	0.454	2.123	0.037	0.644	0.522
H4: TRL4 → BI	0.353	0.725	2.099	0.039	0.162	0.872
H5: TRL5 → BI	1.219	0.226	0.346	0.730	0.423	0.674
H6: TRL6 → BI	0.548	0.585	0.833	0.407	2.030	0.047
H7: TRL7 → BI	0.271	0.787	1.781	0.078	0.155	0.877

where *n* refers to the number of technologies or components of a system, the SRL_i can also be calculated for every single component *i* to identify the normalised values using equation (3.1):

$$\text{Normalised SRL}_i = \text{SRL}_i/m_i \tag{3.2}$$

where m_i refers to the number of integrations of every single component *i* with every other system component. Table 3.8 presents the IRL metric that is characterised as the integration of configuration items (CI) (including hardware and software with the associated risks level) (Long 2011). However, previous metrics were unable to assess the uncertainty due to the integration of a set of technologies comprised of a system. Thus, the SRL was proposed by considering both TRL and IRL as a hybrid solution.

Figure 3.4 shows the average responses for each technology readiness scale level. The responses show that the awareness and initial examples of DTs received more scores from participants. Next, the use of DTs in a pilot facility and testing the full scale of the DT is also confirmed by participants. This may refer to the fact that most sample participants are aware of DTs and examples; however, more use cases are required to be communicated with the industry regarding the compatibility of the DT with other technologies. There is not a strong agreement on higher levels of DT readiness scales.

Table 3.8 Readiness scales and measures for different technologies or systems based on the literature

Individual levels	*IRL (and risks level)*	*SRL to evaluate the whole system (=TRL × IRL)*
1	Configuration item (CI), no interface with the host system (high risk)	Concept refinement (0.10–0.39)
2	CI with some reuse possible (high risk)	Technology development (0.4–0.59)
3	CI with a high degree of reuse possible and incomplete interface requirements (high risk)	System development and demonstration (0.6–0.79)
4	The interface requirements are defined (high risk)	Production and development (0.8–0.89)
5	CI is mature, with no port to new hardware architecture (high risk)	Operations and support (0.90–1.00)
6	CI updated to comply with modified hardware architecture (high risk)	–
7	Changes are required for modifying hardware architecture (med risk)	–
8	CI is compatible with the modified hardware architecture (med risk)	–
9	CI does not need any changes, and operation qualified (low risk)	–
10	CI is embedded and integrated with the system (low risk)	–

Note: TRL: technology readiness levels, SRL: system readiness level, IRL: integration readiness level, MRL: manufacturing readiness level.

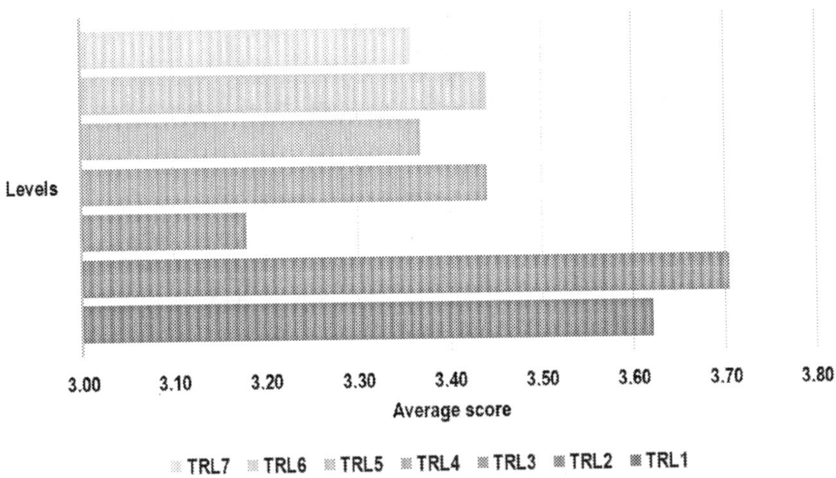

Figure 3.4 An overview of the means and agreements on each TR. (a) Technology readiness levels' mean. (b) The number of participants reported as neutral or agreed to each TR.

3.5 A novel digital twin readiness level scale

Edwards et al. (2017) argue that we are increasingly reliant on machines and the technology that supports, for emancipation, as long as we remain the master of the technology and not a slave to it. While there are multiple readiness scales with seven or nine levels that could be used for technology development (Robinson, Levack et al. 2009), this chapter presents a ten-level scale optimised for assessing DT applications and for use in the construction industry. This is called the digital twin readiness level (DTRL10), as summarised in Table 3.9. This scale provides a solution to Edwards et al.'s (2017) challenge by giving construction professionals the ability to assess technological solutions before blindly adopting them.

The DT maturity assessment scale developed in this chapter reflects the perceptions of practitioners in terms of the current state of the industry. It presents neither the target state for DTs nor a roadmap for the future. It is important that future studies present the DT development map and a set of implementation strategies for different sectors, such as modular construction or infrastructure construction. Notwithstanding the survey limitations, this type of investigation provides valuable insights to research and development units and vendors to address practitioners' concerns and receive insight into critical factors that may affect their intention to use the DT. The method can also be expanded and replicated in different contexts to provide comparative data.

Future studies can test whether industry readiness affects the adoption of DTs and to what extent that depends on the interoperability of associated technologies such as BIM and GIS. Another direction for future research is to examine the interoperability of BIM and GIS with blockchain and data analytics advances in the DT context. Since BIM is widely accepted by practitioners in some construction sectors, a roadmap for the future of BIM in conjunction with DTs should be developed. Also, the cost and schedule of the technology implementation can be further mapped to DTRL to assist construction companies or potential adoptors to have a realistic understanding of the DTs success. The presented SRL and TRL scales are unable to connect the readiness and risks associated with the target market, the required infrastructure and the facilities required for a successful DT implementation in a construction company. Paun (2012) suggested DRL as a matching measure to TRL, enabling the innovators to understand the target market. The DRL is affected by the market push concept and includes some scales to measure the feeling of a missing technology, identification of needs and expected functionality of technology, and sufficient competency of users.

The TRL and other metrics are useful for assessing the technology or demand. However, the key limitation of current metrics is that they cannot assess the producibility of the technology or the feasibility of mass production. Such metrics also ignore the cost in each country or region. The readiness of an organisation, industry, community or country is also dependent on other factors that should be assessed: technology for mass production, cost of adoption, infrastructure to utilise the technology, maintenance cost of the technology, sustainability concerns in the region, availability of relevant skills to utilise the technology. Further investigations should address these limitations.

Table 3.9 A framework for measuring the DTRL considering users' intention to use

Levels and measures	*Development stage*	*Criteria*	*Checklist*
DTRL1: Awareness of the potential of the digital twin (DT) application.	Generate scientific knowledge.	Published scholarly paper underlying the DT concept design.	Have the basic parameters been observed and reported?
DTRL2: An initial example of the DT concept has been introduced.	Identify practical application but with no experimental proof.	Documented the feasibility and benefits of the DT. Defined basic algorithms or coding.	Has the concept been formulated?
DTRL3: Proof of the DT concept has been achieved in the laboratory.	Develop critical properties with but no system integration.	Validated key characteristics or functions of the application.	Has the proof of concept been demonstrated?
DTRL4: The use of DT in a pilot facility has been demonstrated.	Integrate key components of the DT and test interoperability.	Documented the architecture development and the integration test performance.	Has the DT been demonstrated in a laboratory environment?
DTRL5: Proof of the DT concept in a pilot environment has been demonstrated.	Increase the fidelity of components significantly and test end-to-end software system.	Documented the pilot performance.	Has the DT been tested in a simulated construction site?
DTRL6: Proof of the DT application exists, suggesting it will work at full scale.	Demonstrate full-scale DT implementation in a site-like environment.	Documented the test performance and scaling requirements.	Have the DT's feasibility and implementation been demonstrated in a relevant construction environment?
DTRL7: The DT function and its compatibility with other software have been demonstrated in a selected construction site.	Demonstrate an actual prototype in a construction site, and resolve any bugs.	Documented performance with real data analytics.	Has the prototype unit been demonstrated on a relevant construction site?
DTRL8: The DT is proven to work in its final design.	Finalise the development of the DT in a fully integrated way, and complete the verifications.	Documented the performance of the final DT with real data analytics.	Has the DT application been examined – but not yet been operated on the target/selected construction site?

(*Continued*)

Table 3.9 (Continued)

Levels and measures	*Development stage*	*Criteria*	*Checklist*
DTRL9: The actual operation of the DT is evident under the relevant conditions, and benchmarks are available to potential adopters.	Operate the DT successfully under real construction conditions.	Documented the operational results addressing job-site requirements.	Has the DT application been successfully operated in identical construction site situations?
DTRL10: The same DT application was accepted by a wide range of construction stakeholders.	Upgrade and refine the DT to improve its efficiency and functionality.	Report safety of use, with no incident for a protracted time period, and identify failure conditions and downtime.	Has the DT application been certified and operated without incident and with minimal troubleshooting?

Note: From scale 7 onwards, the end users' input should be included.

3.6 Conclusion

This chapter sought to identify critical factors affecting the behavioural intentions of potential users of DTs. A detailed empirical analysis illustrated how six sub-groups of participants perceived the level of the DT and how the technology level may affect the intention of each group to use a DT in their projects.

The information in this chapter is intended to support technology or system developers, particularly those working with DTs in the construction sector, to assess their technology before the acquisition phase. This helps practitioners to understand to what extent technology is mature and estimate the risks of its adoption. Concomitant benefits, therefore, include maximisation of effective decision-making when choosing amongst competing technologies and how these could impact business and project performance; and ameliorating business financial performance by investing in products that generate a tangible return on investment.

The DTRL10 is also useful for scholars as a foundation for framing future experiments to improve the maturity of their technology. At this juncture, it is envisaged that the use of the framework/model developed will provide an invaluable knowledge management control loop to feed back into the model to refine and augment its predictive performance. Moreover, the DTR10 can be used as a guide to assess DT applications in the construction project context.

Ultimately, the DTRL's value is to mitigate DT implementation risks in organisational systems. It offers a common language to assess the maturity of DTs developed by various organisations.

References

Aghimien, D.O., Aigbavboa, C., Edwards, D.J., Mahamadu, A-M., Olomolaiye, P., Onyia, M., and Nash, H. (2020). "A fuzzy synthetic evaluation of the challenges of smart city development in developing countries." *Smart and Sustainable Built Environment*. DOI: https://doi.org/10.1108/SASBE-06-2020-0092

Aheleroff, S., X. Xu, R. Y. Zhong and Y. Lu (2021). "Digital Twin as a Service (DTaaS) in Industry 4.0: An architecture reference model." *Advanced Engineering Informatics* **47**: 101225.

Ahlskog, M., M. Jackson and J. Bruch (2015). *Manufacturing technology readiness assessment*. En paper accepted for POMS 26th Annual Conference.

Ahmed, H., Edwards, D.J., Lai, J.H.K., Roberts, C., Debrah, C., Owusu-Manu, D.G. and Thwala, W.D. (2021). "Post occupancy evaluation of school refurbishment projects: multiple case study in the UK." *Buildings* **11**(4): 169. DOI: https://doi.org/10.3390/buildings11040169

Astuti, M., P. Sudira, F. Mutohhari and M. Nurtanto (2021). "Competency of digital technology: The maturity levels of teachers and students in vocational education in Indonesia." *Journal of Education Technology* **5**(3): 254–262.

Conrow, E. H. (2011). "Estimating technology readiness level coefficients." *Journal of Spacecraft and Rockets* **48**(1): 146–152.

Cronbach, L. J. (1951). "Coefficient alpha and the internal structure of tests." *Psychometrika* **16**(3): 297–334.

Deshmukh Towery, N., E. Machek and A. Thomas (2017). *Technology readiness level guidebook*, Federal Highway Administration.

Edwards, D. J., Pärn, E. A., Love, P. E. D. and El-Gohary, H. (2017). "Machinery, manumission and economic machinations." *Journal of Business Research*, **70**: 391–394. DOI:10.1016/j.jbusres.2016.08.012

Edwards, D.J., Rillie, I., Chileshe, N. Lai, J., Hossieni , M. Reza, and Thwala, W.D. (2020). "A field survey of hand-arm vibration exposure in the UK utilities sector." *Engineering, Construction and Architectural Management*. DOI: https://doi.org/10.1108/ECAM-09-2019-0518

Ellis, J., Edwards, D.J., Thwala, W.D., Ejohwomu, O., Ameyaw, E.E. and Shelbourn, M. (2021). "A case study of a negotiated tender within a small-to-medium construction contractor: Modelling project cost variance." *Buildings* **11**(6): 260. DOI: https://doi.org/10.3390/buildings11060260

Ferreira, C. V., F. L. Biesek and R. K. Scalice (2021). "Product innovation management model based on manufacturing readiness level (MRL), design for manufacturing and assembly (DFMA) and technology readiness level (TRL)." *Journal of the Brazilian Society of Mechanical Sciences and Engineering* **43**(7): 360.

Fisher, L., Edwards, D. J., Pärn, E. A. and Aigbavboa, C. O. (2018). "Building design for people with dementia: a case study of a UK care home." *Facilities* **36**(7/8): 349–368. DOI:10.1108/F-06-2017-0062

Fornell, C. and D. F. Larcker (1981). "Evaluating structural equation models with unobservable variables and measurement error." *Journal of Marketing Research* **18**(1): 39–50.

Gaier, J. R. (2007). *The effects of lunar dust on EVA systems during the Apollo missions* (No. NASA/TM-2005-213610/REV1). Glenn Research Center.

Giancola, P. R. and A. Zeichner (1995). "Construct validity of a competitive reaction-time aggression paradigm." *Aggressive Behavior* **21**(3): 199–204.

Hair Jr, J. F., G. T. M. Hult, C. Ringle and M. Sarstedt (2016). *A primer on partial least squares structural equation modeling (PLS-SEM)*, Sage Publications.

Héder, M. (2017). "From NASA to EU: The evolution of the TRL scale in Public Sector Innovation." *The Innovation Journal* **22**(2): 1–23.

Henseler, J., C. M. Ringle and R. R. Sinkovics (2009). "The use of partial least squares path modeling in international marketing." *New Challenges to International Marketing* **20**: 277–319, Emerald Group Publishing Limited: 277–319. ISSN: 1474-7979. DOI:10.1108/S1474-7979(2009)0000020014

Henseler, J., M. Ringle Christian and M. Sarstedt (2016). "Testing measurement invariance of composites using partial least squares." *International Marketing Review* **33**(3): 405–431.

Hicks, B., A. Larsson, S. Culley and T. Larsson (2009). *A methodology for evaluating technology readiness during product development*. 17th International Conference on Engineering Design (ICED'09) Design Has Never Been This Cool, Stanford University, California, USA, Design Society.

Hulland, J. (1999). "Use of partial least squares (PLS) in strategic management research: A review of four recent studies." *Strategic Management Journal* **20**(2): 195–204.

Ishak, S. S. M. and R. Doheim (2021). "An exploratory study of building information modelling maturity in the construction industry." *International Journal of BIM and Engineering Science* **1**(1): 6–19.

John, G. and T. Reve (1982). "The reliability and validity of key informant data from dyadic relationships in marketing channels." *Journal of Marketing Research* **19**(4): 517–524.

Klar, D., J. Frishammar, V. Roman and D. Hallberg (2016). "A technology readiness level scale for iron and steel industries." *Ironmaking & Steelmaking* **43**(7): 494–499.

Lau, E., Hou, H., Lai, J.H.K., Edwards, D.J. and Chileshe, N. (2021). "User-centric analytic approach to evaluate the performance of sports facilities: a study of swimming pools." *Journal of Building Engineering* **44**. DOI: https://doi.org/10.1016/j.jobe.2021.102951

Long, J. M. (2011). *Integration readiness levels*. 2011 Aerospace Conference, IEEE.

Mankins, J. C. (1995). "Technology readiness levels." White Paper, April 6, 1995.

Newman, C., Edwards, D.J., Martek, I., Lai, J. and Thwala, W.D. (2020). "Industry 4.0 deployment in the construction industry: A bibliometric literature review and UK-based case study." *Smart and Sustainable Built Environment*. DOI: 10.1108/SASBE-02-2020-0016

Pärn, E. A. and Edwards, D. J. (2019). "Cyber threats confronting the digital built environment: Common data environment vulnerabilities and block chain deterrence." *Engineering, Construction and Architectural Management* **26**(2): 245–266. DOI: 10.1108/ECAM-03-2018-0101

Paun, F. (2011). "'Demand Readiness Level' (DRL): A new tool to hybridise market pull and technology push approaches-introspective analysis of the new trends in technology transfer practices." (February 18, 2011). Springer Encyclopedia, Forthcoming, Available at SSRN: https://ssrn.com/abstract=1763679

Paun, F. (2012). The demand readiness level scale as new proposed tool to hybridise market pull with technology push approaches in technology transfer practices. *Technology Transfer in a Global Economy*. D. B. Audretsch, E. E. Lehmann, A. N. Link and A. Starnecker, Springer: 353–366.

Portela, R. M., C. Varsakelis, A. Richelle, N. Giannelos, J. Pence, S. Dessoy and M. von Stosch (2021). "When is an in silico representation a digital twin? A biopharmaceutical industry approach to the digital twin concept." *Digital Twins: Tools and Concepts for Smart Biomanufacturing* **176**: 35–55.

Pylianidis, C., S. Osinga and I. N. Athanasiadis (2021). "Introducing digital twins to agriculture." *Computers and Electronics in Agriculture* **184**: 105942.

Ringle, C. M., S. Wende and J.-M. Becker (2015). "SmartPLS 3. Bönningstedt: SmartPLS." Retrieved 15/07/2016.

Robinson, J., D. Levack, R. Rhodes and T. Chen (2009). *Need for technology maturity of any advanced capability to achieve better life cycle cost*. 45th AIAA/ASME/SAE/ASEE Joint Propulsion Conference & Exhibit. American Institute of Aeronautics and Astronautics.

Rybicka, J., A. Tiwari and G. A. Leeke (2016). "Technology readiness level assessment of composites recycling technologies." *Journal of Cleaner Production* **112**: 1001–1012.

Samad, M. E. S., S. Shirowzhan and M. Loosemore (2021). "Information asymmetries between vendors and customers in the advanced construction technology diffusion process." *Construction Innovation* **21**(4): 857–874.

Sauser, B., D. Verma, J. Ramirez-Marquez and R. Gove (2006). *From TRL to SRL: The concept of systems readiness levels*. Conference on Systems Engineering Research, Los Angeles, CA, Citeseer.

Sepasgozar, S. M. (2020a). "Digital technology utilisation decisions for facilitating the implementation of Industry 4.0 technologies." *Construction Innovation* **21**(3): 476–489.

Sepasgozar, S. M. E. (2020b). Digital twin and cities. *The Palgrave encyclopedia of urban and regional futures*, Robert C. Brears. Springer International Publishing: 1–6.

Sepasgozar, S. M. E. (2022). "Immersive on-the-job training module development and modeling users' behavior using parametric multi-group analysis: A modified educational technology acceptance model." *Technology in Society* **68**: 101921.

Sepasgozar, S. M., A. A. Khan, K. Smith, J. G. Romero, X. Shen, S. Shirowzhan, H. Li and F. Tahmasebinia (2023). "BIM and digital twin for developing convergence technologies as future of digital construction." *Buildings* **13**(2): 441.

Shirowzhan, S., S. Lim and J. Trinder (2016). "Enhanced autocorrelation-based algorithms for filtering airborne lidar data over urban areas." *Journal of Surveying Engineering* **142**(2): 04015008.

Shirowzhan, S., S. M. E. Sepasgozar, D. J. Edwards, H. Li and C. Wang (2020). "BIM compatibility and its differentiation with interoperability challenges as an innovation factor." *Automation in Construction* **112**: 103086.

Song, J. and F. M. Zahedi (2005). "A theoretical approach to web design in E-commerce: A belief reinforcement model." *Management Science* **51**(8): 1219–1235.

Straub, J. (2015). "In search of technology readiness level (TRL) 10." *Aerospace Science and Technology* **46**: 312–320.

Straub, J. (2021). *Evaluating the use of Technology Readiness Levels (TRLs) for cybersecurity systems*. 2021 IEEE International Systems Conference (SysCon), IEEE.

Sullivan, B. P., E. Arias Nava, S. Desai, J. Sole, M. Rossi, L. Ramundo and S. Terzi (2021). "Defining Maritime 4.0: Reconciling principles, elements and characteristics to support maritime vessel digitalisation." *IET Collaborative Intelligent Manufacturing* **3**(1): 23–36.

Uflewska, O., T. Wong and M. Ward (2017). *Development of technology maturity framework in managing manufacturing improvement for innovation providers*. 24th Innovation and Product Development Management Conference, Reykjavik University in Reykjavik.

Visner, M., S. Shirowzhan and C. Pettit (2021). "Spatial analysis, interactive visualisation and GIS-based dashboard for monitoring spatio-temporal changes of hotspots of bushfires over 100 years in New South Wales, Australia." *Buildings* **11**(2): 37.

Wu, C., B. Xu, C. Mao and X. Li (2017). "Overview of BIM maturity measurement tools." *Journal of Information Technology in Construction (ITcon)* **22**(3): 34–62.

Zutin, G. C., G. F. Barbosa, P. C. de Barros, E. B. Tiburtino, F. L. F. Kawano and S. B. Shiki (2022). "Readiness levels of Industry 4.0 technologies applied to aircraft manufacturing—A review, challenges and trends." *The International Journal of Advanced Manufacturing Technology* **120**(1–2): 927–943.

4 Digital twin adoption modelling incorporating job relevance, usefulness, and relative advantage

An empirical investigation

Marco Mura, Samad M. E. Sepasgozar, Sara Shirowzhan, Alberto De Marco and Michael J. Ostwald

4.1 Introduction

The construction sector is gradually being reshaped by the introduction of virtual reality (VR), mixed reality (MR) solutions, building information modelling (BIM) and associated tools, and digital twins (DTs). As the sector becomes aware of the benefits of combining DTs with the existing construction frameworks and tools, the demand for DT technology is growing. NASA provided an early definition of DT technology as integrating physical and digital entities, which may be multi-scale with the probabilistic simulation of a system (Shafto, Conroy et al. 2010). As an integrated multi-discipline, multi-scale, probabilistic simulation of a complex product, a DT uses the best available physical models and sensor updates to mirror its corresponding twin's life (Cheng, Zhang et al. 2020). In the manufacturing sector, DTs simulate a production system that is characterised by its synchronisation with the real system. In this way, the DT is used for forecasting and optimising the behaviour of the production system over its life cycle and in real-time (Cimino, Negri et al. 2019).

A DT typically has five core components, two from the physical world (sensors, actuators) and three from the digital world (data, integration, analytics):

- *Sensors*: create signals to allow the DT to capture operational and environmental data coming from the physical process in the real world (Parrott and Warshaw 2007).
- *Data*: real-world signals are aggregated and combined with data from the enterprise (bill of materials (BOM), enterprise system, and design specifications) (Parrott and Warshaw 2007).
- *Integration*: technology allows communication and combining data between the physical and digital worlds.
- *Analytics*: techniques are used to analyse the data through algorithmic simulations and visualisation routines that the DT uses to produce insights (Khosravanian and Aadnøy 2022).

DOI: 10.1201/9781003507000-4

- *Actuators*: if the DT identifies intolerable deviations from the optimal conditions or weak points in the process that can be improved, the DT responds by way of actuators, subject to human intervention, which shapes a change in the physical process.

Industry policymakers' acceptance of the "Industry 4.0" concept increases the need to understand practitioners' behaviours in terms of technology adoption and explore under what circumstances the acceptance rate will be expedited. Industry 4.0 was initially promoted in 2011 to foster the digitalisation of manufacturing and many other sectors (Savastano, Amendola et al. 2019). Industry 4.0 in the AEC industry is called "Construction 4.0" or "Building 4.0." Its essence is the digitalisation and automation of the AEC industry (García de Soto, Agustí-Juan et al. 2019). Industry 4.0 embraces different technologies and associated paradigms, including radio frequency identification (RFID), enterprise resource planning (ERP), Internet of Things (IoT), cyber-physical systems (CPS), Internet of Services (IoS), artificial intelligence (AI), machine learning (ML), big data, smart production and cloud-based manufacturing amongst others (Lu 2017; Yang and Chou 2019). Most of these technologies are fundamental for developing DT applications. The Industry 4.0 concept refers to a new industrial stage in which manufacturing operations systems and information and communication technologies (ICTs), are highly integrated (Dalenogare, Benitez et al. 2018). Construction 4.0 and its implementation improve this process and take decisions almost in real-time, supported by data provided by many new technologies and equipment like prefabrication, automation, VR, drones, sensors, and robots (Yang and Chou 2019).

The purpose of the research presented in this chapter is to gain an understanding of critical factors that will affect a practitioner's decision to use a DT application. This chapter extends the body of knowledge in the field by first identifying DT acceptance factors and, second, setting the identified factors into specific contexts by comparing different groups of practitioners in terms of experience or company size. The first contribution is crucial since previous studies in the construction contexts examined the user's intention for single technologies. DT applications, including a set of tools and maintaining real-time data exchange, have not been fully developed, and their adoption has not been investigated in various contexts. The latter comparisons are even more critical since previous studies mainly examined general practitioners in the construction context. The influence of the selected factors on users' intentions to use a DT application by different groups of practitioners has not been covered. By addressing these two knowledge gaps, this research contributes to the theoretical extension of technology acceptance by identifying a set of new variables specified to DT applications that influence practitioners' adoption decisions.

4.2 Overview of digital twin applications

A DT generally refers to the digital counterpart of an object (Sepasgozar 2020). The virtual entity relies on the collected sensed and transmitted data of the IoT infrastructure and can elaborate them in a connected and smart way (Lu, Xie et al. 2020). Compared to the total number of publications regarding DT technology in all industries, the construction industry seems to be at an early stage in developing practical DT applications (Lu, Liu et al. 2019).

Previous studies have identified multiple DT applications in different industries, with many in the aerospace industry, where the DT concept was born. For instance, according to (Glaessgen and Stargel 2012), DT technology can predict every event and condition during the twin flying experience, simulating what the physical twin will experience in a real flight situation. Glaessgen and Stargel state that having access to those types of data and simulations allows the DT to play a central role in terms of fleet management and sustainment. Moreover, the DT can constantly monitor the physical twin, providing vital feedback, allowing the DT to be useful beyond the standalone simulation process, enhancing the decision-making phase to reduce risk during the in-flight changes and in anomalous situations.

As an example of a DT, general electric (GE) developed a holistic DT model to improve the overall performance of a power plant (Khalyasmaa, Stepanova et al. 2023). This DT can remotely monitor the local weather forecast, every plant's energy demand peak time, and their thermal efficiency and performance. Once the data are collected, the DT runs simulations, indicating the most efficient configuration and optimising the whole process. The outcomes include: improved operational flexibility, reducing plant start-up time by up to 50%; improved asset performance management and its reliability, reducing unplanned outages to save up to $150MM/year and delivering up to $5MM of additional MWh; optimisation of the inventory management strategy, which reflects in a cost maintenance reduction up to 10%. GE is also using a DT application in the aviation industry, focusing on jet engines. This DT can monitor the engine through sensor data when exposed to extreme operating conditions while suggesting design and performance improvement for the next prototypes. In a third example, BP used DT technology to design and verify a physical project within the oil and gas industry (Zborowski 2018; Wanasinghe, Wroblewski et al. 2020). Having the opportunity to monitor a piece of remote equipment with high-fidelity accuracy can support improved safety conditions for workers and enhanced predictive maintenance strategy, typical issues within the oil and gas industry.

Predicting the life of aircrafts and their parts is a complex process involving a combination of different physics models: the computational fluid dynamics model (CFD), the structural dynamics model (SDM), the thermodynamic model, the fatigue cracking model (FCM), and the stress analysis model (SAM), to mention few of the possible models. Before DTs, these different models were treated separately. According to (Tuegel, Ingraffea et al. 2011), the DT elaborates the FCM, SAM, SDM and other possible models holistically. Moreover, the DT gives engineers access to extensive information and data, facilitating the database visualisation of both the structural and the physical model, enabling a more efficient configuration control of the aircraft and better maintenance decisions, which will reflect in an improved overall management of the aircraft.

In the manufacturing industry, DT applications have supported the development of smart manufacturing processes. According to (Qi and Tao 2018), DTs can play a central role either in the design phase or in the creation of a smart workshop or factory where they support usage monitoring and smart maintenance and repair processes. During the design phase, the DT monitors client expectations, factoring in real-world constraints. Due to the continuous flow of data, the DT can detect defects during the design phase and suggest changes, verifying their feasibility and running high-fidelity digital simulations instead of real-world simulations, saving money and time.

Tao, Cheng et al. (2018) similarly underline the importance of DTs during all parts of the design process (conceptual design, schematic design, detailed design, and virtual verification). During conceptual design, the DT helps stakeholders visualise all the information and data combined, such as customer satisfaction and investment plans. In the detailed design stage, designers refine the project in every detail, transforming the concept into a form that can be manufactured. This stage requires several simulations performed by the DT to ensure the performance and characteristics required. The last stage of the manufacturing design process is known as product verification. However, instead of being performed in the real-world environment, with a DT, it can be performed virtually, reducing production waste and time taken during the testing and verification phase.

According to Tao, Cheng et al. (2018), DT technology is the key to creating an efficient smart workshop. Combining all aspects – geometrical models, equipment, material, environment, optimal resource allocation, and capabilities – it can simulate manufacturing strategies to choose the most suitable one. Moreover, the ultra-high synchronisation between the physical process and its virtual counterpart with its auto-updating, allows the process to be monitored in real time and, if necessary, be modified to improve performance. The monitoring tool also enables smart maintenance, repair, and overhaul (MRO) processes. Having real-time access to the health and performance of the physical twin, combined with the simulation tool that considers the updated state of the physical asset, historical data and real-world environment, improves predictions of the actual health and remaining life of an asset. This supports proactive maintenance to avoid faults and unexpected equipment downtime, which would affect the whole process (Tuegel, Ingraffea et al. 2011).

Some of the most popular applications of DTs are in prognostics and health management (PHM) (Tao, Zhang et al. 2019). DTs provide the opportunity for a fault diagnostics and lifetime prediction strategy that holistically takes into account every aspect of the environment (e.g., physics, behaviour, simulations, data, geometry), not just historical data or static physical data. When considering the whole-life cycle asset management in the construction industry, existing tools like BIM are not always enough, especially in the operation and maintenance phases. DT integrates AI, ML, and data analytics to create dynamic digital models that are able to learn and update the status of the physical counterpart from multiple sources (El Jazzar, Piskernik et al. 2020; Lu, Xie et al. 2020). Research into the effectiveness of DTs in these fields emphasises the ways they can enhance asset performance and productivity (Macchi, Roda et al. 2018).

4.3 Proposed conceptual model

There are two overarching approaches in the literature to examine technology adoption in the architectural, engineering and construction contexts. The first overarching approach is to utilise smaller samples of approximately 30 participants or less and case study evidence identifying critical factors (motivators, drivers, and barriers) that may influence the technology adoption process. These investigations are often based on interviews for qualitative analysis or basic statistics. The second approach is to develop a model based on previous structured models, such as the technology acceptance model (TAM) or the unified theory of acceptance and use of technology (UTAUT). In this second approach, a set of key constructs are chosen to establish the model and identify how these may lead an individual to accept a new technology based on structural modelling techniques. While the first approach is more exploratory, the latter is mainly confirmatory based on modifications or replications of the generic theoretical models in different contexts. Generic models such as TAM or UTAUT are widely used in the literature and often highlight "perceived usefulness" and "ease of use." Recent versions of TAM, such as TAM 2 and 3, offer four variables for measuring the usefulness or ease of use of a general technology such as a smartphone. However, there is no strong empirical evidence offering a set of measures for DTs in the construction sector. To address the lack of specified measures of usefulness for DT applications, a set of variables are proposed in this section as measures of usefulness.

Based on the relevant literature, some variables are designed to measure the usefulness of a DT. A DT can be useful if it enhances the construction equipment operation process by quickly verifying its feasibility (PU1), reduces operational risks by simulating heavy equipment in a construction site (PU2), enhances the equipment PHM process during construction (PU3), decreases equipment downtime (PU4), and is useful for assessing the equipment performance (PU5). These variables are thought to potentially affect the job relevance (JB) and relative advantage (RA) of the DT application.

The literature suggests that using DTs can ensure lower operational risks, due to the possibility of running simulations in advance of the entire operation process, according to the operating conditions of the physical, real-world environments (Qi and Tao 2018). Due to its functions, a DT can forecast and verify the design process and its virtual verification. The simulations can find the defects of the design in the virtual world and make rapid changes if required (Qi and Tao 2018; Cheng, Zhang et al. 2020). One of the main benefits of DTs is related to equipment PHM as well as downtime. Predictive maintenance extends the life of heavy equipment, while prognostics allows predicting the time at which the machinery will no longer perform its functions, meeting the desired performance levels (Qi and Tao 2018; Cheng, Zhang et al. 2020). The DT also allows monitoring of equipment parameters during the operational phase, and it integrates multiple dynamic simulations in different environments and increases productivity. A DT continually records data and operating parameters, comparing the actual product response to the anticipated product response in a specific scenario (Qi and Tao 2018; Cimino, Negri et al. 2019). This suggests that the equipment performance assessment is also a potential measure of perceived usefulness.

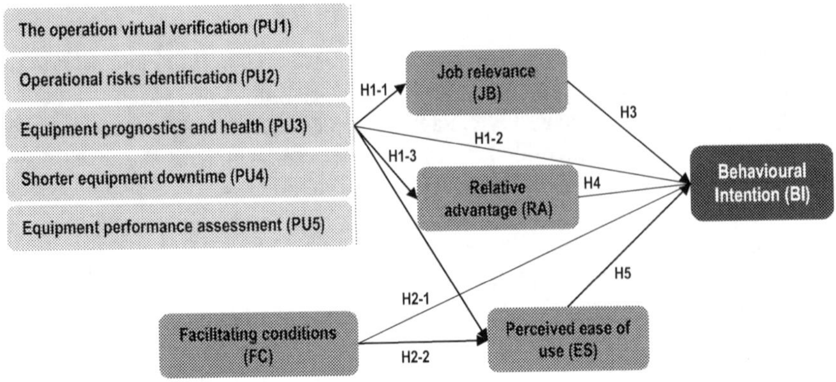

Figure 4.1 The structure of the proposed digital twin based on the TAM and UTAUT hybrid.

Perceived usefulness is related to an individual's belief that a particular system enhances their job performance (Davis 1989; Venkatesh and Davis 2000). This construct is part of the TAM. Venkatesh states that there is a positive relation between perceived usefulness and adoption intention because of the benefits that the innovation brings to the users. Consequently, the hypotheses are that perceived usefulness has a positive effect on JB, RA, and users' adoption intention (H1–1 to 1–3).

Figure 4.1 shows the proposed conceptual model of this research. The two main constructs are perceived ease of use and perceived usefulness. The constructs and key factors were identified through a literature review of individual and social constructs influencing technology acceptance. Table 4.1 shows the list of constructs, with their related definition from the literature, hypothesis, and measures.

According to (Venkatesh, Thong et al. 2012), a relationship between facilitating conditions and technology use may affect the adoption intention. Facilitating conditions are related to factors in the environment that observers agree will make an act easy to do (Thompson, Higgins et al. 1991). This construct is part of UTAUT (Venkatesh, Thong et al. 2012). It suggests that the technology user would find the system easier to use in positive facilitating conditions. Consequently, the hypotheses are that facilitating conditions have a positive effect on the intention to use and a positive effect on the DT's perceived ease of use.

Perceived ease of use is associated with an individual's perception that using the system would be easy and effortless, and it comes from the TAM (Davis 1989; Venkatesh and Davis 2000). The definition suggests that users' adoption intention will be higher if a new technology is easy to use. Consequently, the hypothesis is that perceived ease of use positively affects user intention.

The relative construct advantage is part of the innovation diffusion theory (IDT) (Venkatesh, Morris et al. 2003) and represents the user's perception that the new system is perceived as being superior to its predecessor (Moore and Benbasat

Table 4.1 Hypothetical relationship of constructs and relevant measures

Construct	*Definition and resources*	*Hypothesis*	*Measures*
Perceived usefulness (PU)	The perception that using DT may improve relevant task performance (Venkatesh and Davis 2000). The degree to which a person believes that using a particular system would enhance his or her job performance (Davis 1989).	Perceived usefulness has a positive effect on DT user's adoption intention.	PU1, PU2, PU3, PU4, PU5
Perceived ease of use (ES)	The perception that the use of the technology will be easy and effortless (Venkatesh and Davis 2000). The degree to which a person believes that using a system would be free of effort (Davis 1989).	Perceived ease of use has a positive effect on DT user's adoption intention.	ES1, ES2
Job relevance (JR)	The degree to which an individual believes that the target system. is applicable to his or her job (Venkatesh and Davis 2000).	Job relevance has a positive effect on DT's perceived usefulness.	JB1, JB2
Relative advantage (RA)	Perception of technology as superior to its predecessors (Rogers 2010). The degree to which using innovation is perceived as being better than using its precursor (Moore and Benbasat 1991).	Relative advantage has a positive effect on the DT's perceived usefulness.	RA1, RA2
Facilitating conditions – technology (FC)	Objective factors in the environment that observes agree to make an act easy to do, including the provision of computer support (Thompson, Higgins et al. 1991).	Facilitating conditions has a positive effect on DT's perceived ease of use.	FC1
Behavioural intention (BI)	Probability, likelihood, and willingness to adopt a DT (Song and Zahedi 2005).		BI1, BI2, BI3

1991; Rogers 2010). Consequently, the hypothesis is that relative advantage positively affects the DT's perceived usefulness.

Job relevance concerns the belief that the target system is applicable to the job required. It is part of the TAM. Venkatesh and Bala (2008) found evidence that job relevance is one factor that captures the influence of cognitive processes on perceived usefulness. Moreover, they state that the higher the job relevance, the higher its effect on perceived usefulness. Consequently, the hypothesis is that job relevance positively affects DT's perceived usefulness.

4.4 Research method

A structured questionnaire is designed to collect data for the present research. The questionnaire is composed of a set of questions measuring key constructs of the proposed DT AM. The questions are based on a five-point Likert scale, from 1 = strongly disagree to 5 = strongly agree. A pilot survey was conducted to ensure all questions, which are designed based on academic literature, are also clear and meaningful to practitioners. Two weeks were dedicated to conducting the pilot survey, involving five experts in the field who provided vital feedback to make the questionnaire the clearest possible. Most of the feedback addressed the length of the questions, which could have led to having a complicated survey to complete. Moreover, feedback from the pilot survey was useful, in terms of clearness of the survey, which helped not to have ambiguous questions. For instance, feedback for the question "Assuming I had access to the digital twin, I am motivated to use it" was reviewed and modified to "Assuming I had access to the digital twin, I would be motivated to use it." This is to clarify that it does not refer to the actual intention to use DT, but the possible future adoption.

Following human ethics approval (HC190984), data was collected by distributing a survey to the target participants selected using social media. Participants were identified via LinkedIn by searching three different keywords within the "People" section. The three keywords are: (1) digital twin, construction industry, (2) BIM, construction industry, and (3) digital technologies, construction industry. Then, every profile was checked to verify if it complied with the inclusion/exclusion criteria. The criteria were: (i) the organisations involved could be public, private, or universities; (ii) their core business must be the construction industry, such as an architectural firm, general construction firm, facility management firm, or similar businesses; (iii) practitioners must have a managerial role in their organisations or technical expertise; (iv) practitioners must have some experience in the construction industry and in implementing digital technologies such as BIM. Since people with less than three years of experience were also asked to answer the survey, an ad hoc question was necessary to evaluate their experience. The question "How many years have you been using one or more of the technologies listed above?" allowed the team to elaborate on data separately and interpret them based on the participants' experience. The research is limited to Australian construction companies. Nine hundred and thirty-seven profiles were identified as eligible. The survey was conducted using the Qualtrics platform, and the link provided to participants led to it. A total of 937 surveys were distributed, and 234 were returned, representing a 24.97% response rate). Among the submitted responses, 192 were valid for the analysis of this chapter because they answered the questions that are relevant to technology acceptance. Structural equation modelling was used to analyse the data. Figure 4.2 provides an overview of the participants' profiles.

Figure 4.2 An overview of the participant's profile. (a) Participants' main business, (b) participants' expertise, (c) the percentage of participants who are aware of potential digital twin applications.

4.5 Analysis and results

The proposed DTAM includes 15 variables, as shown in Table 4.2. These variables are used to test the relationship among the key constructs of DTAM. The verified relationship is important to confirm or modify the DTAM. The partial least squares (PLS) analysis method suggested by Ringle, Wende et al. (2015) was used for testing the relationships among constructs of the proposed DTAM. This section presents the results of the required tests for validating the model, including convergent validity and discriminant validity (John and Reve 1982; Giancola and Zeichner 1995). These tests examine relationships between variables and the correctness of their reliability. The minimum observed variance inflation factor (VIF) varies from 1.000 to 3.026, which is well below the recommended value of 5.0 (Henseler, Ringle et al. 2009; Hair Jr, Hult et al. 2016). Table 4.2 shows descriptive statistics of the original sample, *t*- and *p*-values, loadings factor, and the outer VIF values.

The reliability of the scale and internal consistency of constructs were evaluated by computing two main measures Cronbach's alpha and composite reliability (CR). Table 4.3 shows that Cronbach's alpha ranges from 0.606 to 0.889, and the CR values vary from 0.835 to 0.942, and all are acceptable since they are higher than the threshold of 0.6 (Fornell and Larcker 1981), although ES has a marginal value. Table 4.3 also shows sufficient evidence of convergent validity since the average variance extracted (AVE) values vary from 0.717 to 0.890, which is greater than 0.5 (Fornell and Larcker 1981). The PLS algorithm results show that BI, ES, JB, and RA have values of 0.21 to 0.522 for the total variances explained in their

Table 4.2 **The DTAM variables and their descriptive analysis *t*- and *p*-values and loadings and VIF**

ID	*Variables*	*Observed max*	*STDEV*	*Cramér-von Mises test statistic*	*Cramér-von Mises p-value*	*Loading*	*Outer VIF*
BI1	Assuming I had access to the DT, I would be motivated to use it	2.790	0.439	1.212	0.000	0.899	2.786
BI2	Given that I had access to the DT, I predict that I would use it	1.246	0.394	1.152	0.000	0.919	3.026
BI3	I plan to use the DT within the next 12 months	1.558	0.665	1.516	0.000	0.747	1.401
ES1	I believe that it would be easy to get DT to do what I want it to do	1.345	0.523	1.798	0.000	0.853	1.233
ES2	I believe that DT would be easier to use compared to the current practices of using different tools for equipment monitoring	1.922	0.541	1.798	0.000	0.841	1.233
FC1	DT would be compatible with our current systems	0.000	0.000	1.530	0.000	1.000	1.000
JB1	In my job, the use of DTs would be important	0.993	0.327	5.143	0.000	0.945	2.552
JB2	The use of DTs would be pertinent to my various job-related tasks	1.720	0.336	5.143	0.000	0.942	2.552
PU1	I believe that DT would enhance the construction equipment operation process by quickly verifying its feasibility	1.464	0.535	1.127	0.000	0.845	2.706
PU2	I believe that DT would reduce operational risks by simulating heavy equipment in a construction site	1.720	0.546	1.966	0.000	0.838	2.404
PU3	I believe that DT would help to enhance the equipment prognostics and health management process during construction	1.264	0.499	1.437	0.000	0.867	2.784
PU4	I believe that DT would decrease equipment downtime	1.247	0.580	1.838	0.000	0.815	2.404
PU5	I believe that DT would be useful for assessing the equipment performance	1.791	0.604	2.232	0.000	0.797	2.170
RA1	Using DT would enable me to accomplish tasks more quickly	1.281	0.376	6.287	0.000	0.927	2.185
RA2	Using DT would improve the quality of the work I do	1.547	0.350	6.287	0.000	0.937	2.185

Note: Number of valid observations for each variable = 193, STDEV: standard deviation, VIF: variance inflation factor.

Table 4.3 Construct reliability and validity outcomes

LV	*Cronbach's alpha*	*Rho_A*	*CR*	*AVE*	*Q^2 (=1 – SSE/SSO)*	*R square*	*R square adjusted*
BI	0.817	0.835	0.893	0.737	0.368	0.522	0.509
ES	0.606	0.606	0.835	0.717	0.128	0.210	0.202
FC	0.876	0.877	0.942	0.890	0.000		
JB	0.889	0.893	0.919	0.693	0.351	0.397	0.394
PU	0.848	0.851	0.929	0.868	0.000		
RA	0.817	0.835	0.893	0.737	0.320	0.374	0.371

Note: CR: composite reliability, AVE: average variance extracted.

Table 4.4 Discriminant validity of constructs based on Fornell–Larcker criteria

Constructs	*BI*	*ES*	*FC*	*JB*	*PU*	*RA*
BI	0.858					
ES	0.314	0.847				
FC	0.401	0.372	1.000			
JB	0.655	0.379	0.347	0.943		
PU	0.599	0.388	0.378	0.630	0.833	
RA	0.636	0.354	0.441	0.747	0.611	0.932

Note: Diagonals: the square roots of average variance extracted (AVE).

respective constructs, which are all well above 2% as a recommended threshold. This shows that BI, as the main construct, has the highest explained variance.

Table 4.4 shows that Fornell and Larcker's (1981) criteria for testing independence and the level of correlation were satisfied. The square roots of the AVEs are superior to the corresponding inter-variable correlations. Each construct falls in the range of 0.717 to 0.890, well above 1.5% of its total variance explained. Table 4.4 shows the diagonal cells of the discriminant validity matrix, including the correlation between latent variables ranging from 0.847 to 1.000, which are much greater than the recommended cut-off of 0.7. Table 4.4 shows that all variables have individual reliability coefficients superior to 0.4, which is above the minimum recommended acceptable level for exploratory studies (Hulland 1999). The results of the tests show that the convergent and discriminant validity of the measures and the divergence validity of the proposed model were satisfied.

4.5.1 *The confirmation of the structural model of VTAM*

The Student's *t*-test was utilised to measure the significance of the coefficient for the proposed paths through the bootstrapping technique. To do so, 5,000 subsamples were created using the original data as a replacement, where the analysis was initially run with 500 subsamples. The *t*-test shows the real difference between the groups. The value of the *t*-test is considered significant if it is equal to or greater than 1.96 (Hair Jr, Hult et al. 2016). Table 4.5 shows that the results of f^2 vary

Table 4.5 Summary of the hypotheses of the digital twin adoption model, including estimated path coefficients with *t*-value

Path	*O*	*M*	*STDEV*	t-*Values*	p-*Values*	*Inner VIF*	f^2	*Accept*
ES → BI	−0.014	−0.012	0.061	0.223	0.824	1.297	0.000	Not supported
FC → BI	0.115	0.113	0.067	1.713	0.087	1.346	0.021	Not supported
FC → ES	0.263	0.262	0.086	3.071	0.002	1.166	0.075	Supported
JB → BI	0.315	0.313	0.090	3.486	0.000	2.583	0.080	Supported
PU → BI	0.232	0.231	0.086	2.697	0.007	1.878	0.060	Supported
PU → ES	0.288	0.289	0.080	3.601	0.000	1.166	0.090	Supported
PU → JB	0.630	0.630	0.046	13.594	0.000	1.000	0.660	Supported
PU → RA	0.611	0.611	0.056	10.875	0.000	1.000	0.597	Supported
RA → BI	0.212	0.217	0.099	2.148	0.032	2.604	0.036	Supported

Note: H: hypotheses, M: sample mean, STDEV: standard deviation, path coefficients: direct effect, f^2: effect size.

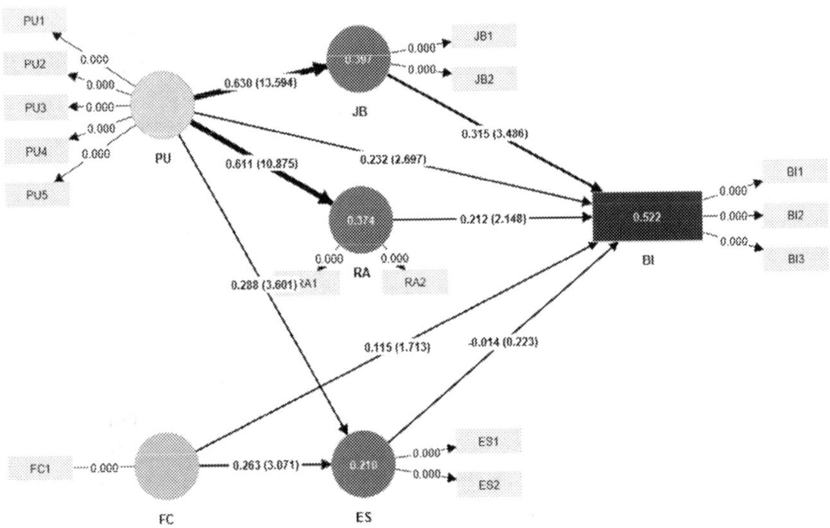

Figure 4.3 Schematic of digital twin adoption model, including relative values of path coefficient showed by the thickness of lines and *t*-values for the inner and the *p*-values of outer relationships.

from 0.000 to 0.660 for supported relationships, representing acceptable predictive relevance by having average to high explanation power. The f^2 values for ES → BI are weak since it is less than 0.02, suggested by Hair Jr, Hult et al. (2016). The Stone-Geisser's q^2 was computed to validate the redundancy by running the "blindfolding" procedure with an omission distance (OD) of seven ($D = 7$). The model could predict the point indicators classified as endogenous measures (refer to Table 4.5). Figure 4.3 shows the results of evaluating the DT adoption model structure, including testing the hypothetical relationships. The path coefficients and t-statistics of inner and outer relationships are shown in Figure 4.3.

Table 4.5 shows the outcome of structural modelling tests, including a two-tailed t-test from the bootstrap. Table 4.5 shows that inner VIF is well below 5, ranging from 1.000 to 2.583, indicating that collinearity is not a problem in this analysis. Overall, the outcomes of the tests support all hypotheses, excluding two of them. The paths ES → BI and FC → BI are not supported, so ES and FC cannot directly be used to predict BI.

4.5.2 DTAM validation by comparing for various groups of participants

Additional tests were applied to explore whether the path coefficients differed significantly between various groups of participants. Two tests are multi-group parametric and Welch–Satterthwaite, which were performed four times on the tested structural model using various groups within the whole sample. These groups are based on (1) company size, (2) participants' experience, (3) the innovative category of the company, and (4) the experience of using digital technology.

As a requirement of conducting the multi-group assessment, the permutation algorithm was applied to different groups A, B, C, and D (Henseler, Ringle Christian et al. 2016). The variation of permutation p-value for each group is as follows: A: 0.032–0.780, B: 0.116–0.952, C: 0.014–0.789, and D: 0.031–0.875. The differences between RA → BI (Group A), FC → ES (Group C), and RA → BI (Group C) may be significant. Besides these two coefficients, the premutation analyses of all other group variables are apparently non-significant (p-value > 0.05). Tables 4.6 and 4.7 show the parametric and Welch–Satterthwaite test results.

The parametric results of t and p-values show that there are significant differences between the groups on nine paths in Groups A, B, C, and D. In Group A, the significant differences are between RA → BI, $t = 2.144$; and $p = 0.035$. In Group C, the significant differences are between FC → ES, RA → BI, $t = 3.158$, 2.189 and 2.006; and $p = 0.032$ and 0.049, respectively. However, more tests and discussions are made for each moderator (e.g., company size and participants' experience), as follows.

The company size was used as the first moderator to test the model by two sub-samples of participants from smaller and larger companies. Participants from companies with employees less than 200 are considered smaller to medium-sized companies and are called Group 1. Participants from companies with employees of more than 199 are considered large-sized companies and are called Group 2. The number of participants from Groups 1 and 2 is 100 and 92, respectively. These two groups were used to make Group A for the statistical companion.

Figure 4.4 shows the outcome of the comparison, and differences appear between RA and BI and between the FC and ES. However, this does not necessarily mean the differences are statistically significant. Consequently, the three-step tests recommended by the invariance of composite models were used, and the criteria were met. For example, Step 3 was met since the p-values for means and variances of all constructs of the model vary between 0.165 and 0.945 (>0.05), showing they are

Table 4.6 Multi-group parametric assessment results of p-values for all groups

Paths	*Group A*		*Group B*		*Groups C*		*Group D*	
	t-*Value*	p-*Value*	t-*Value*	p-*Value*	t-*Value*	p-*Value*	t-*Value*	p-*Value*
ES → BI	1.937	0.054	0.060	0.952	0.803	0.423	0.157	0.875
FC → BI	1.286	0.200	0.453	0.651	0.268	0.789	1.021	0.308
FC → ES	1.118	0.265	0.545	0.586	2.474	0.014	2.168	0.031
JB → BI	0.853	0.395	0.109	0.913	1.054	0.293	0.550	0.583
PU → BI	1.449	0.149	1.579	0.116	1.804	0.073	0.983	0.327
PU → ES	0.279	0.780	1.461	0.146	0.417	0.677	0.662	0.509
PU → JB	1.787	0.075	0.621	0.535	1.114	0.267	1.113	0.267
PU → RA	1.613	0.108	0.504	0.615	1.152	0.251	1.230	0.220
RA → BI	2.157	0.032	1.179	0.240	2.009	0.046	0.477	0.634

Table 4.7 Welch–Satterthwaite results of p-values for all groups

Paths	*Group A*		*Group B*		*Group C*		*Group D*	
	t-*Value*	p-*Value*	t-*Value*	p-*Value*	t-*Value*	p-*Value*	t-*Value*	p-*Value*
ES → BI	1.928	0.057	0.059	0.953	0.835	0.407	0.149	0.882
FC → BI	1.295	0.198	0.422	0.674	0.265	0.791	1.085	0.283
FC → ES	1.103	0.273	0.570	0.570	2.189	0.032	1.616	0.114
JB → BI	0.848	0.399	0.101	0.920	1.087	0.281	0.561	0.578
PU → BI	1.452	0.150	1.685	0.095	1.620	0.110	1.129	0.264
PU → ES	0.278	0.781	1.519	0.132	0.355	0.724	0.438	0.663
PU → JB	1.762	0.081	0.583	0.562	1.105	0.273	0.816	0.419
PU → RA	1.591	0.115	0.479	0.633	1.118	0.268	0.858	0.396
RA → BI	2.144	0.035	1.139	0.258	2.006	0.049	0.528	0.600

the same or consistent. Thus, the path coefficient of permutation tests was checked. It shows that p-values vary from 0.070 to 0.797 (>0.05), showing no significant difference between the relationships in Groups 1 and 2. Further statistics of t-tests and p-values are shown as Group A in Tables 4.6 and 4.7. The outcome shows that the variation of parametric and Welch–Satterthwaite p-value for RA → BI may differ for large companies, with $p = 0.032$ and 0.035, respectively.

The experience of the employee, in terms of the number of years a participant has been in the industry, was used as the second moderator to test the model by two sub-samples of participants from smaller and larger companies. Participants with three years or less experience of working in the industry were considered as less experienced and called Group 14. Participants with four years or more of experience working in the industry are considered as more experienced and called Group 24. The number of participants from Groups 14 and 24 is 41 and 151, respectively. These two groups were used to make Group D for the statistical companion.

Figure 4.5 shows the outcome of the comparison, and differences appear between RA and BI and between the FC and ES. However, this does not necessarily mean the differences are statistically significant. Consequently, the three-step

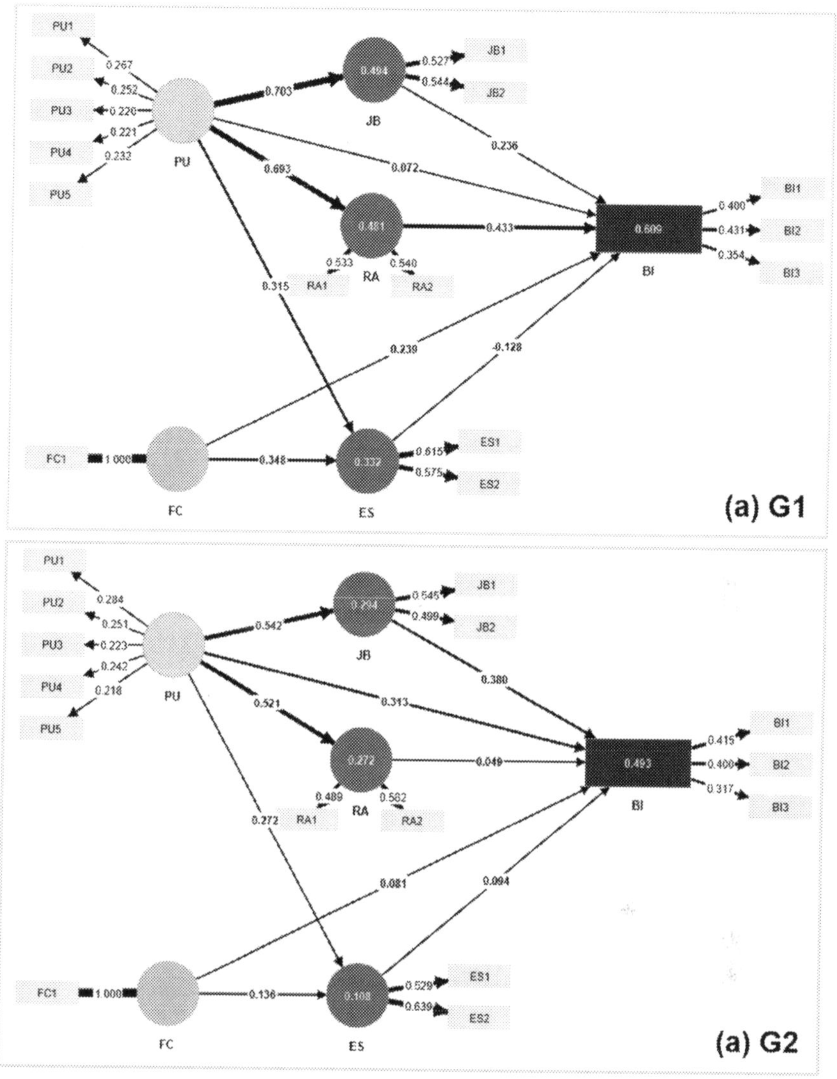

Figure 4.4 Comparing the group bootstrapping results for Groups 1 and 2 based on company size.

tests recommended by the invariance of composite models were used, and the criteria were met. For example, Step 3 was met since the p-values for means and variances of all constructs of the model vary between 0.380 and 0.965, 0.166 and 951, respectively, showing they are >0.05 and consistent. Thus, the path coefficient of permutation tests was checked. It shows that p-values vary from 0.244 to 0.957 (>0.05), showing no significant difference between the relationships in Groups 12 and 22. Further statistics of t-tests and p-values are shown as Group A in Tables 4.6

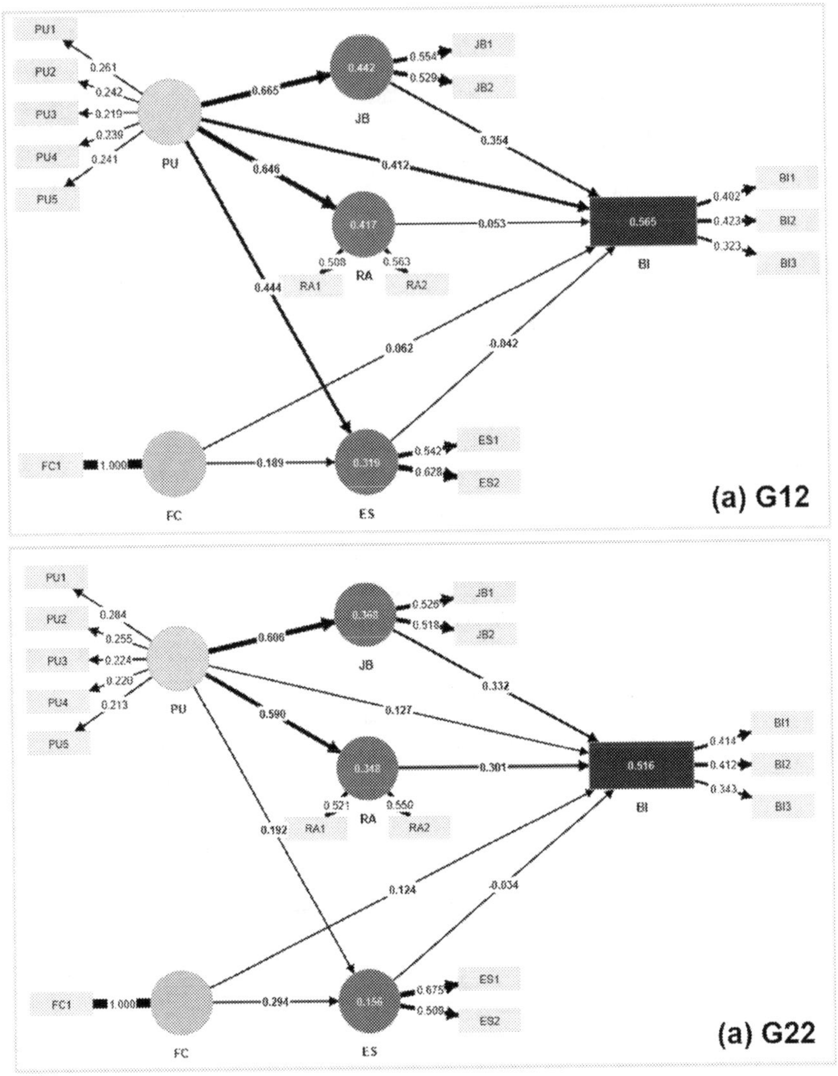

Figure 4.5 Comparing the group bootstrapping results for Groups 12 and 22 based on participants' experience.

and 4.7. The outcome shows that the parametric and Welch–Satterthwaite *p*-values are not significant.

The innovation adoption category of a company was used as the third moderator to test the model by two sub-samples of participants from smaller and larger companies. Table 4.8 shows the definitions used to describe each company, and participants categorised themselves as Group 13 or 23 based on the definition.

Figure 4.6 shows the comparison outcome, and differences appear in some paths, such as between RA and BI. However, this does not necessarily mean the

Table 4.8 Group C, including participants, categorised their companies as early or later adopters

Group C	*Definitions used for the investigation*	*Number of complete answers*
Group 13	We are early adopters because we adopt new digital technologies as soon as they come to the market, probably 84% earlier than similar companies in our industry	74
Group 23	We are conservative, so late adopters of new digital technology probably adopt it earlier than 50% or wait until 50% of similar companies in our industry have adopted the technology	118

differences are statistically significant. Consequently, the three-step tests recommended by the invariance of composite models were used, and the criteria were met. For example, Step 3(a) was partially met since the *p*-values for means of BI and ES constructs of the model are 0.094 and 0.231, respectively, showing they are <0.05 and not significant. Step 3(b) was partially met since the *p*-values for variances of FC construct is 0.019, showing it is <0.05 and not significant. The path coefficient of the group bootstrap shows that the relationship between FC and ES is 0.04 (<0.05). However, the bootstrapping results for all relationships were checked to identify their consistency. The outcome of the parametric and Welch–Satterthwaite *p*-values for path coeffects of FC to ES and RA to BI are significant (<0.05). Since impartial invariance is established, the bootstrapping results were examined, and it shows that the *p*-values for the following relationships are not consistence, so they are not invariant: FC → ES, JB → BI, PU → BI, PU → ES, and RA → BI. These tests show that the DT adoption pattern may differ for various categories of adaptors, or the constructs may affect their intention to use DTs differently.

The number of years a participant used technology is used as the fourth moderator to test the model by two sub-samples of participants with the experience of using a type of IoT, visualisation technology, augmented or VR, AI, big data analysis tools, cyber-physical systems (CPS), and cloud computing. Participants with three years or less experience using any of these technologies were considered less experienced and called Group 14. Participants with four years or more experience using digital technology in their business were considered more experienced and called Group 24. The number of participants from Groups 14 and 24 is 41 and 151, respectively. These two groups were used to make Group D for the statistical companion.

Figure 4.7 shows the outcome of the comparison, and differences appear in some paths, such as between PU and RA with BI. To check these, the three-step tests recommended by the invariance of composite models were used, and the criteria were met. For example, Step 3(a) was partially met since the *p*-values for means of JB and ES constructs of the model are 0.021 and 0.036, respectively,

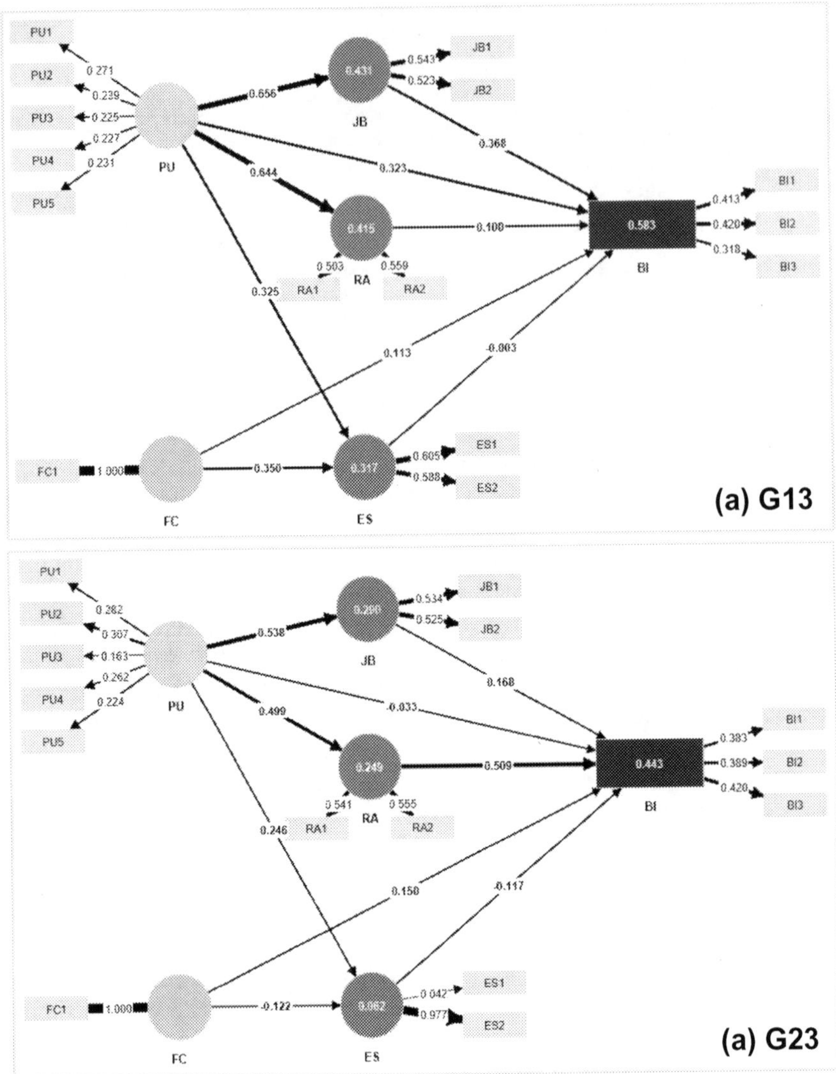

Figure 4.6 Comparing the group bootstrapping results for Groups 13 and 23 based on the innovation adoption category.

showing they are <0.05 and not significant. Step 3(b) was fully met since the p-values for variances of all constructs are >0.05. The path coefficient of the group bootstrap shows that the relationship between FC → ES is 0.04 (<0.05). The bootstrapping results for all relationships were then checked. The outcome of the parametric and Welch–Satterthwaite p-values for path coeffects of FC → ES and RA to BI are significant (<0.05). Their experience of using another digital technology does not provide strong evidence of being different in some paths of the structural model.

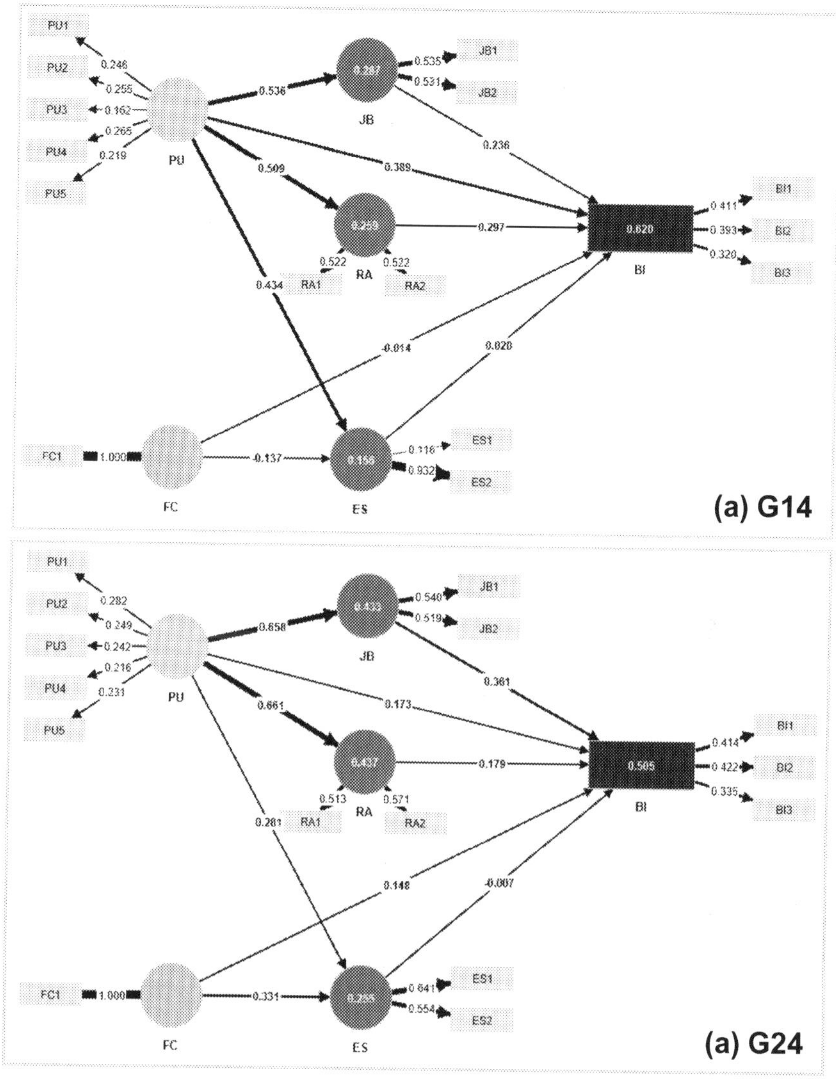

Figure 4.7 Comparing the multiple-group assessment bootstrapping results for Groups 14 and 24: USE.

4.6 Conclusion

This chapter aimed to develop an adjusted technology adoption model based on the current general theoretical TAM, to be used for the prediction of DT utilisation in construction contexts. A total of 937 links to the survey were distributed, and 234 participants responded, representing a 24.97% response rate. Among the submitted responses, 192 were valid for the analysis carried out in this chapter. A significant portion of the participants, around 76%, expressed their intention to use a DT application if it is made available to them (refer to Figure 4.8).

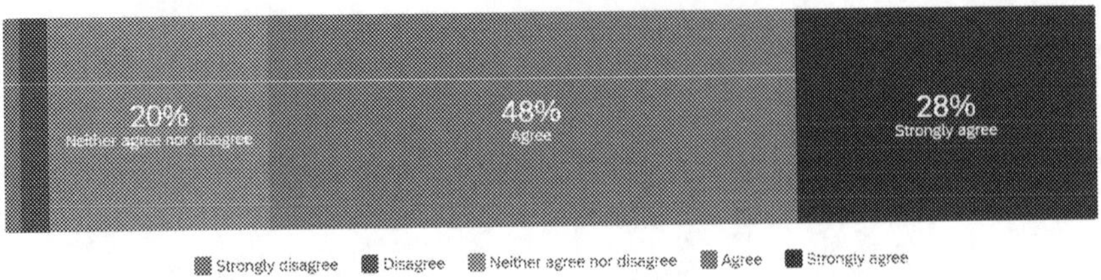

Figure 4.8 The percentage of participants who wish to use digital twins if available to them.

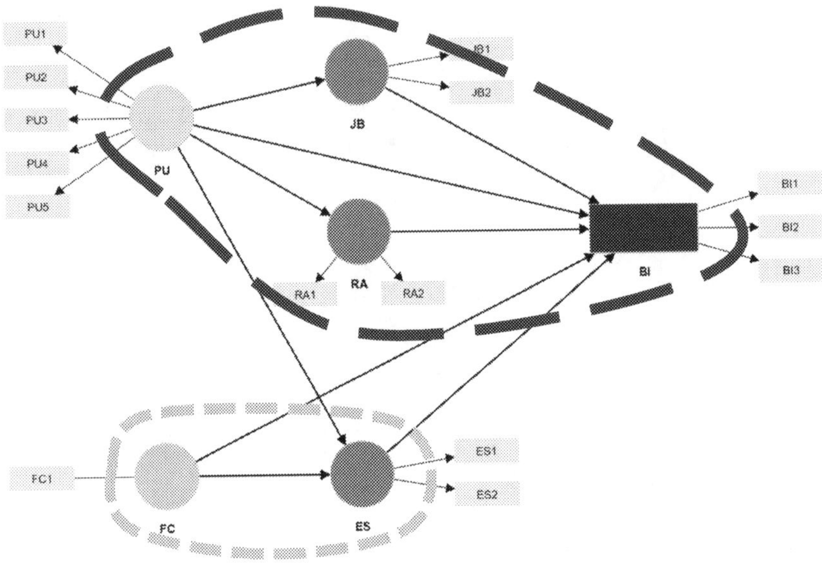

Figure 4.9 A simplified schematic of DTAM. Influential factors for digital twin utilisation (top) and the relationship between facilitating conditions and ease of use (bottom).

The chapter revealed the relationship between various constructs of the DTAM, explaining the predictors of the users' intention to utilise a DT application. Figure 4.9 shows two sets of constructs associated with DT utilisation. The perceived usefulness, job relevance, and relative advantage of the DT application depicted at the top of the figure have been identified as strong predictors of the users' intention to utilise the application in the construction context. However, the application's facilitating conditions and ease of use have not been found to impact its utilisation significantly. While usefulness has been known as a strong factor in other sectors, it refers to a different set of variables for a DT in the construction context. The investigation results indicate that the perceived usefulness of a DT application is a critical factor influencing its utilisation. The chapter specified that in the construction context, the application's ability to provide virtual verification of operational processes, identify operational risks, diagnose equipment issues and health, control asset breakdowns, and assist in asset performance assessment had been found to be particularly influential in promoting the adoption of DT technology.

The findings of this empirical investigation have significant implications for the adoption and utilisation of DT applications in construction organisations or projects. Specifically, the results demonstrate that practitioners are likely to prioritise a DT application's usefulness and job relevance over ease of use and facilitating conditions. Suppose a DT can assist in improving or controlling the challenging and complex construction processes. In that case, practitioners are willing to utilise the application, even if it is not easy to operate.

This theoretical DTAM offers valuable insights and can help technology managers in various companies when developing or utilising DT applications. By prioritising the usefulness and job relevance of DT applications, companies can create applications that meet the needs of DT users in the construction industry. Furthermore, DTAM can be adjusted and localised based on different contexts and can be examined using various sets of participants.

Despite this study's statistical and sampling limitations, the chapter provides strong evidence of predicting factors for DT adoption, with the participation of a large sample of practitioners. The findings of this study can serve as a valuable resource for companies and researchers interested in the adoption and utilisation of DT applications in the construction industry.

References

Cheng, J., H. Zhang, F. Tao and C.-F. Juang (2020). "DT-II: Digital twin enhanced Industrial Internet reference framework towards smart manufacturing." *Robotics and Computer-Integrated Manufacturing* **62**: 101881.

Cimino, C., E. Negri and L. Fumagalli (2019). "Review of digital twin applications in manufacturing." *Computers in Industry* **113**: 103130.

Dalenogare, L. S., G. B. Benitez, N. F. Ayala and A. G. Frank (2018). "The expected contribution of Industry 4.0 technologies for industrial performance." *International Journal of Production Economics* **204**: 383–394.

Davis, F. D. (1989). "Perceived usefulness, perceived ease of use, and user acceptance of information technology." *MIS Quarterly: Management Information Systems* **13**(3): 319–339.

El Jazzar, M., M. Piskernik and H. Nassereddine (2020). *Digital twin in construction: An empirical analysis*. EG-ICE 2020 Workshop on Intelligent Computing in Engineering, Proceedings, Universitätsverlag der TU Berlin.

Fornell, C. and D. F. Larcker (1981). "Evaluating structural equation models with unobservable variables and measurement error." *Journal of Marketing Research* **18**(1): 39–50.

García de Soto, B., I. Agustí-Juan, S. Joss and J. Hunhevicz (2019). "Implications of Construction 4.0 to the workforce and organizational structures." *International Journal of Construction Management* **22**(2): 205–217.

Giancola, P. R. and A. Zeichner (1995). "Construct validity of a competitive reaction-time aggression paradigm." *Aggressive Behavior* **21**(3): 199–204.

Glaessgen, E. and D. Stargel (2012). *The digital twin paradigm for future NASA and U.S. air force vehicles*. American Institute of Aeronautics and Astronautics.

Hair Jr, J. F., G. T. M. Hult, C. Ringle and M. Sarstedt (2016). *A primer on partial least squares structural equation modeling (PLS-SEM)*, Sage Publications.

Henseler, J., C. M. Ringle and R. R. Sinkovics (2009). The use of partial least squares path modeling in international marketing. *New challenges to international marketing*, Tamer Cavusgil, Rudolf R. Sinkovics, Pervez N. Ghauri. Emerald Group Publishing Limited: 277–319.

Henseler, J., M. Ringle Christian and M. Sarstedt (2016). "Testing measurement invariance of composites using partial least squares." *International Marketing Review* **33**(3): 405–431.

Hulland, J. (1999). "Use of partial least squares (PLS) in strategic management research: A review of four recent studies." *Strategic Management Journal* **20**(2): 195–204.

John, G. and T. Reve (1982). "The reliability and validity of key informant data from dyadic relationships in marketing channels." *Journal of Marketing Research* **19**(4): 517–524.

Khalyasmaa, A. I., A. I. Stepanova, S. A. Eroshenko and P. V. Matrenin (2023). "Review of the digital twin technology applications for electrical equipment lifecycle management." *Mathematics* **11**(6): 1315.

Khosravanian, R. and B. S. Aadnøy (2022). Chapter One - Introduction to digital twin, automation and real-time centers. *Methods for petroleum well optimization*. R. Khosravanian and B. S. Aadnøy, Gulf Professional Publishing: 1–30.

Lu, Y. (2017). "Industry 4.0: A survey on technologies, applications and open research issues." *Journal of Industrial Information Integration* **6**: 1–10.

Lu, Y., C. Liu, K. Wang, H. Huang and X. Xu (2019). "Digital twin-driven smart manufacturing: Connotation, reference model, applications and research issues." *Robotics and Computer-Integrated Manufacturing* **61**: 101837.

Lu, Q., X. Xie, J. Heaton, A. K. Parlikad and J. Schooling (2020). From BIM towards digital twin: Strategy and future development for smart asset management. *Studies in Computational Intelligence* **853**: 392–404.

Macchi, M., I. Roda, E. Negri and L. Fumagalli (2018). "Exploring the role of digital twin for asset lifecycle management." *IFAC-PapersOnLine* **51**(11): 790–795.

Moore, G. C. and I. Benbasat (1991). "Development of an instrument to measure the perceptions of adopting an information technology innovation." *Information Systems Research* **2**(3): 192–222.

Parrott, A. and L. Warshaw. (2007). "Industry 4.0 and the digital twin." https://www2.deloitte.com/us/en/insights/focus/industry-4-0/digital-twin-technology-smart-factory.html.

Qi, Q. and F. Tao (2018). "Digital twin and big data towards smart manufacturing and Industry 4.0: 360 degree comparison." *IEEE Access* **6**: 3585–3593.

Ringle, C. M., S. Wende and J.-M. Becker (2015). "SmartPLS 3. Bönningstedt: SmartPLS." Retrieved 15/07/2016.

Rogers, E. M. (2010). *Diffusion of innovations*, Simon and Schuster.

Savastano, M., C. Amendola, B. Bellini and F. D'Ascenzo (2019). "Contextual impacts on industrial processes brought by the digital transformation of manufacturing: A systematic review." *Sustainability* **11**(3): 891.

Sepasgozar, S. M. E. (2020). Digital twin and cities. *The Palgrave encyclopedia of urban and regional futures*, Robert Brears (Editor-in-Chief). Springer International Publishing: 1–6.

Shafto, M., M. Conroy, R. Doyle, E. Glaessgen, C. Kemp, J. LeMoigne and L. Wang (2010). "Draft modeling, simulation, information technology & processing roadmap." *Technology Area* **11**:1–32.

Song, J. and F. M. Zahedi (2005). "A theoretical approach to web design in E-commerce: A belief reinforcement model." *Management Science* **51**(8): 1219–1235.

Tao, F., J. Cheng, Q. Qi, M. Zhang, H. Zhang and F. Sui (2018). "Digital twin-driven product design, manufacturing and service with big data." *The International Journal of Advanced Manufacturing Technology* **94**(9): 3563–3576.

Tao, F., H. Zhang, A. Liu and A. Y. C. Nee (2019). "Digital twin in industry: State-of-the-art." *IEEE Transactions on Industrial Informatics* **15**(4): 2405–2415.

Thompson, R. L., C. A. Higgins and J. M. Howell (1991). "Personal computing: Toward a conceptual model of utilization." *MIS Quarterly: Management Information Systems* **15**(1): 125–142.

Tuegel, E., A. Ingraffea, T. Eason and S. Spottswood (2011). "Reengineering aircraft structural life prediction using a digital twin." *International Journal of Aerospace Engineering* **2011**: 1–14.

Venkatesh, V. and F. Davis (2000). "A theoretical extension of the technology acceptance model: Four longitudinal field studies." *Management Science* **46**: 186–204.

Venkatesh, V., M. Morris, G. Davis and F. Davis (2003). "User acceptance of information technology: Toward a unified view." *MIS Quarterly* **27**: 425–478.

Venkatesh, V. and H. Bala (2008). "Technology acceptance model 3 and a research agenda on interventions." *Decision Sciences* **39**: 273–315.

Venkatesh, V., J. Y. L. Thong and X. Xu (2012). "Consumer acceptance and use of information technology: extending the unified theory of acceptance and use of technology." *MIS Quarterly* **36**(1): 157–178.

Wanasinghe, T. R., L. Wroblewski, B. K. Petersen, R. G. Gosine, L. A. James, O. De Silva, G. K. Mann and P. J. Warrian (2020). "Digital twin for the oil and gas industry: Overview, research trends, opportunities, and challenges." *IEEE Access* **8**: 104175–104197.

Yang, J. B. and H. Y. Chou (2019). "Subjective benefit evaluation model for immature BIM-enabled stakeholders." *Automation in Construction* **106**: 102908.

Zborowski, M. (2018). "Finding meaning, application for the much-discussed "digital twin." *Journal of Petroleum Technology* **70**(06): 26–32.

5 Immersive virtual environments and digital twin applications for education and training

Trends in construction, mining, and urban planning studies

Samad M. E. Sepasgozar, Ayaz Ahmad Khan, Sara Shirowzhan, Juan Sebastian Garzon Romero, Christopher Pettit, Chengguo Zhang, Joung Oh and Ruiyu Liang

5.1 Introduction

One of the significant applications of digital twin and immersive technologies (DTIT) is in education, whether this relates to training, learning, or education itself (Radianti et al., 2020). This involves workers/employees at construction sites or in offices, and students in universities, and is aimed at improving knowledge of the overall mechanism of a project (Bouska & Heralova, 2019). Advanced technologies offer a realistic scenario-based environment to obtain knowledge by doing, increasing skill and awareness of a certain task (Dai et al., 2023; Qiu et al., 2023). The ability to track the learner's progress based on interactivity through DTIT is far more effective than purely physical training. Education via DTIT promotes the ability to track the learner's progress based on the situation faced, which helps to give targeted support to the individual. Also, integrating game-based learning with DTIT environments helps retain knowledge for more time, enhancing the experience through the development of skills and abilities (Kandi et al., 2020). The recent and anticipated growth of the DTIT market is significant, with the global DT market value in the year 2020 thought to be USD $3.1 billion and expected to reach USD $48.2 billion by the year 2026 at a substantial compound annual growth rate (CAGR) of 58% (Markets, 2020, September). On the other hand, the global market of immersive techniques is expected to reach USD $814.7 billion at a CAGR of 63.01% (Zion Market Research, 2019, February 21).

DTIT are the core technologies that support learning, teaching, and education processes (Nikolaev et al., 2018), and are preferably the underlying concepts to derive the ideology of Education 4.0 (E-4.0) (Gomerova et al., 2021). The concept of E-4.0 leapfrogged over the backdrop of Industrial Revolution 4.0 (IR-4.0), which integrates digital technologies in design, production, and commercialisation to achieve enhanced products and services. Although IR-4.0 possesses the potential

DOI: 10.1201/9781003507000-5

to transform various industries, the lack of digital training and education is a serious hindrance to its development and operations (Hariharasudan & Kot, 2018). Thus comes the concept of E-4.0, which integrates these technological applications into learning and training.

Based on "learning by doing," E-4.0 can be defined as a new training and educational paradigm, enhancing digital competencies among professionals/workers and students/educators in the sector (David, 2018; Madni, 2019). Although early research classified E-4.0 communication modes, such as webinars, online courses, digital media, and YouTube learning, a significant impact is achieved by employing virtual learning environments, which provide knowledge more intuitively (Almeida & Simoes, 2019). Also, as the fields of architecture, engineering, and construction (AEC), mining, and urban planning (UP) are more application-based, DTIT could provide significant support for the process of E-4.0, eventually addressing the potential of IR-4.0 (Joe David, 2018; Song et al., 2020). Therefore, there is a need to conduct reviews of the current applications to identify future paths for the successful development of E-4.0 and IR-4.0 in these sectors.

The AEC field has seen significant research focused on education and training through DTIT (Alizadehsalehi et al., 2021; Chacón, 2021). This approach can provide realistic scenarios for professionals and students, ultimately enhancing their skills and knowledge. By building upon previous educational practices, the industry aims to equip individuals with the necessary abilities to excel in their projects (Rafsanjani & Nabizadeh, 2021; Yitmen et al., 2021). The ease of simulating dangerous activities and tasks like fall hazard scenarios at the site is a major advantage of DTIT learning, presenting these concepts based on real activities to make it easier to learn (Eiris et al., 2018). Training workers with various machinery involving operations like cranes, bulldozers, and pile boring machines, among others, can be better displayed to the concerned user (Sepasgozar, 2020). The cost reduction in travelling for workers is also a factor that makes immersive learning useful. Workers and trainees can also practice with demanding activities and heavy machinery beforehand, like crane maneuvers, scale wind turbines, and 3D visualisation.

On the other hand, students can learn and understand complex architectural and structural analysis by understanding spatial arrangements and complex elements. The simulated environment created through DTIT also allows students to judge and verify human cognitive behaviour, which aids in designing more sustainable buildings (Khan et al., 2021). Challenging courses like structural engineering and analysis can become tedious to students, bringing difficulties in understanding concepts and techniques. However, through immersive environments, the comprehension of gained skills remains for a long time through the improved visualisation processes they offer. Prior training in probable circumstances can also reduce the likeness of other risks, like electrical hazards.

Similarly, the mining industry has capitalised on this technology to enhance the education related to various aspects of the sector. The work in the mining industry is considered threatening in nature. Therefore, new solutions provided by the implementation of DTIT escalate the overall education in this sector (Grabowski & Jankowski, 2015). Occupational health and safety, acquisition and practice of correct behaviours, and monitoring of the operations in the mining industry can be facilitated by the implementation of DTIT.

Regarding urban planning, city planning, and city analytics, DTIT is used as a medium to construct interactive public displays for deeper participation. People become engaged with the urban planning of a city, for example, through visualisation platforms (Dembski et al., 2020), pedestrian movements, playful public participation, bringing motivation for people to adopt new planning ideas and concepts (Du et al., 2019; Fonseca et al., 2017; Kent et al., 2019). In addition, integrating the potential of geographic information systems (GIS) with DTIT can enhance policymakers' and urban planners' knowledge of a city or an urban space at a miniature scale (Zhenjiang Shen, 2012).

Among different immersive techniques such as virtual reality (VR), augmented reality (AR), and mixed reality (MR), VR has been the most used technique in various sectors. The application of VR at a wide scale is because of the seamless environment it provides for education. Also, creating VR solutions and ideas is much more effective and productive than AR or MR. Although some sectors of the construction industry have employed DTIT applications, knowledge of their applicability and processes is scattered in the fields of AEC, mining, and UP. The need for virtuality, especially during the widespread Covid pandemic that occurred around the world, justifies the development of new remote and virtual learning (Lev Rassudov, 2020). It created the need to provide a state-of-the-art DTIT to guide this process in current work and the future paths to follow. As mentioned earlier, anticipated growth is expected for these fields. Therefore, this chapter will present ongoing applications and additional suggestions that would contribute to the realisation of the growth likely in this field.

Following this backdrop, this chapter aims to review the various studies of DTIT in education and training in three different sectors: (1) AEC disciplines, (2) mining, and (3) urban planning. Since there are different practices applying digital tools in education, learning from those carried out in similar sectors is necessary to identify overlaps and gaps in the relevant literature. This can be achieved by reviewing publications in various disciplines. Previous reviews are mainly focused on general VR applications without a focus on specific sectors (González-Zamar & Abad-Segura, 2020), or they focus on one specific discipline without considering other relevant practices (Chen et al., 2020). Thus, there is a need to conduct a review of the application of DTIT in education. The clear objectives are (I) to classify and summarise modules developed or used in these three sectors, (II) to identify the state-of-the-art and its limitations, and (III) to offer future research directions.

The next section in this study describes the methodology employed for the chapter, followed by the results of critical content analysis in Section 5.3. Further, Section 5.4 is a discussion based on the findings, comprising the state of play and likely future directions. Lastly, the conclusions of this study are deliberated in Section 5.5.

5.2 Materials and methods

Initially, a literature review was conducted systematically using the Scopus database following the preferred reporting process for systematic reviews and the meta-analysis PRISMA protocol. The keywords for this systematic study were as follows and conducted separately for all three fields:

(TITLE-ABS-KEY ("digital twin" OR "virtual reality" OR "augmented reality" OR "mixed reality" OR "immersive") AND TITLE-ABS-KEY (architecture OR construction OR building) AND TITLE-ABS-KEY (education OR learn* OR teach* OR train*))

(TITLE-ABS-KEY ("digital twin" OR "virtual reality" OR "augmented reality" OR "mixed reality" OR "immersive technology*") AND TITLE-ABS-KEY (Mining) AND TITLE-ABS-KEY (education OR learn* OR teach* OR train*))

(TITLE-ABS-KEY ("digital twin" OR "virtual reality" OR "augmented reality" OR "mixed reality" OR "immersive technology*") AND TITLE-ABS-KEY ("City analytics" OR "city planning" OR "Urban planning") AND TITLE-ABS-KEY (education OR learn* OR teach* OR train*))

5.2.1 Selection criteria

The initial search within the three domains of AEC, mining, and urban planning, returned 6,652, 741, and 108 articles, respectively. The screening criteria were based on limiting the years of search to "2011 to 2022" since the chapter intends to review practices of immersive techniques in the last decade. Practices before 2011 are unlikely to be useful for this chapter since the technologies they utilised are outdated and most likely unavailable in the market. The document type is termed "article," the source type is denoted as "journal," and the language selection of "English" is used since journal publications should present practices with validation tests and be peer-reviewed before publication. After this screening process, 87, 13, and 5 articles remained for the three respective domains, as shown in Figure 5.1.

In addition to this, the Google Scholar database is also used for domains 2 and 3 as the number of articles identified by Scopus was few. To obtain relevant results for domains 2 and 3, Google Scholar was utilised to search the first ten pages, with the search being sorted by relevance. This yielded a total of 26 and 20 additional articles, respectively. The search strategy and selection for the study are shown in Figure 5.1, and the final number of articles for each domain is shown in Table 5.1. A few publications from recognised conferences were also taken to deliver in-depth knowledge of domains 2 and 3.

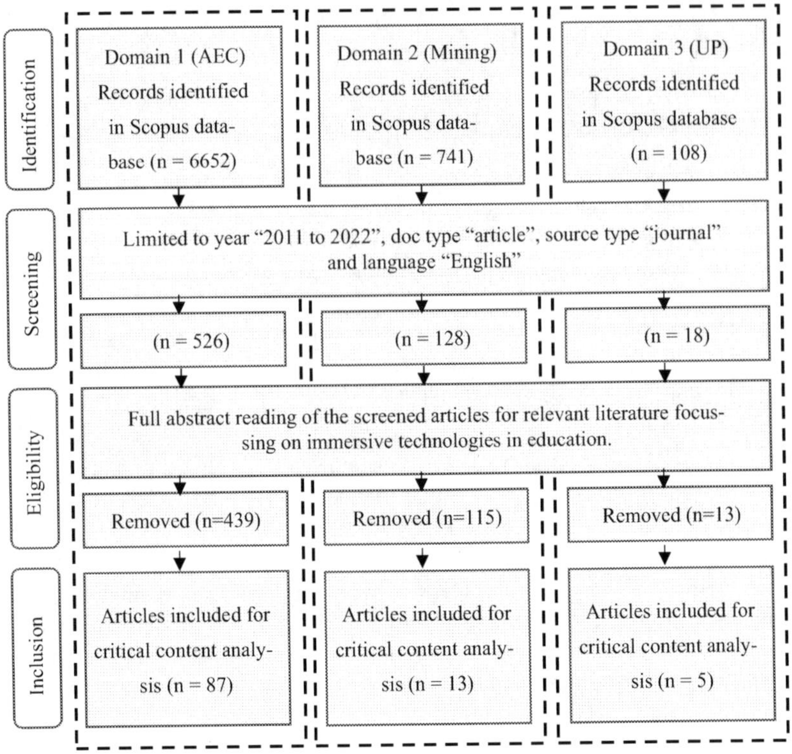

Figure 5.1 Research method including the search strategy and selection criteria.

Table 5.1 Final included articles for critical content analysis

Domain	*No. of publications*		*Total*
	Scopus	*Google Scholar*	
AEC	87	–	87
Mining	13	26	39
Urban planning	5	20	26
Total			152

5.2.2 Data extraction and critical content analysis

To ensure that the data is structured in a useful manner, all publications from each domain were thoroughly reviewed and analysed. This rigorous process involved the critical examination of all 152 publications, with the goal of effectively synthesising the qualitative data.

5.3 Results

This section provides the outcome of the systematic review in the selected disciplines: (1) AEC disciplines, (2) mining, and (3) urban planning. Some concepts are difficult to convey through studies, such as what is happening underground during construction, what topography means and how it affects planning, and how to improve earning about information systems and visualising data.

5.3.1 AEC industry

The surge of digital technologies in the AEC industry is not only restricted to providing satisfaction in terms of need; rather, they have become functional representatives in education and training purposes. This study synthesises 87 publications relevant to DTIT in AEC; the outcomes are described below.

DTIT includes DTs and various tools and facilities to provide an immersive learning environment, such as VR, AR, and MR (Alizadehsalehi et al., 2020). DT can be defined as a virtual, synthetic, or digital replica of a built asset, system, or process and can be static or dynamic in nature. Although its main application is anticipated in simulations, predictions, and decision-making, DT-based training and education are significant areas for promoting active learning using tangible and intangible assets.

The use of immersive techniques in the AEC industry has been widely researched and reported in the past. For example, studies have found that VR can be an effective tool for construction workers to identify and assess risks in a project through training. The increasing use of VR in research programs at universities is also helping to automate the AEC industry, particularly with the advent of IR-4.0.

The latest research demonstrates the advantages of incorporating AR and MR technologies in education for specific disciplines. In a recent study, researchers conducted a workshop that included a lecture and hands-on experience for construction students in an AR/MR environment, resulting in improved learning outcomes as demonstrated by the qualitative results. Another study utilised AR/MR technologies as a digital teaching method for AEC courses in universities (Khan et al., 2021). The study combined five digital technologies using AR/MR techniques and DT to enhance the literature by showcasing the effectiveness of construction courses (Khan et al., 2021). Table 5.2 presents the top highly cited publications in education and training for the AEC sector, focusing on addressing critical safety issues in construction and training workers to create safe work environments.

Few practical applications developed by the authors describe the usage of DTIT in various domains for AEC. Figure 5.2 shows four main types of practice for communicating the data and visualising equipment or buildings with students. These practices are categorised into augmented and simulation technologies that intend to present external objects or data analytics. The first practice refers to lifelogging, such as dashboarding and data mining, which can be used to show the dynamics of city areas. The second practice refers to the augmented reality of designed objects. The third practice refers to mirroring a city in 3D. The fourth practice refers to VR

Table 5.2 Top highly cited articles in the AEC discipline

Article focus	*Method and description of the study*	*Outcomes of the study*
Safety training in an immersive virtual environment (Sacks et al., 2013)	Safety training and learning of the workers in the VR environment. Two groups were divided – one for VR training and the other for conventional training, and the results were compared.	The effectiveness of VR training was found to be more effective in some tasks. However, more intuitive VR training is required for general site safety.
VR/AR applications in safety (Li et al., 2018)	A comprehensive review on VR/AR for safety in construction highlighting various prototypes, products, and training paradigms.	A taxonomy is generated comprising technology, applications, scenarios, and evaluation methods in VR/AR for safety.
A framework for and the use of visualisation systems (Park & Kim, 2013)	A construction safety framework is developed integrating BIM, AR, and gaming related to safety management and visualisation.	The framework is tested on a case study for various accident scenarios and results in better risk recognition capacity.
Location tracking and data visualisation for training ironworkers (Teizer et al., 2013)	A novel approach is integrating real-time location tracking and VR for steel erection risk tasks in the workers' indoor training centre.	The results show a significant understanding of risks for workers, along with better effectiveness of VR training.
Gaming technologies for safety improvements (Guo et al., 2012)	Game-based safety training in the construction plant and equipment is carried out to identify the best approach to learning preventive actions.	The developed game-based platform enhanced the knowledge abilities of workers in construction plant operations.
Develop a social VR-based system (Le et al., 2015)	A conceptual framework for assessing the university students' knowledge of safety and health based on online VR.	Online modules of different scenarios for construction safety were conducted, and results showed improved knowledge in users.
VR for visualising a bridge cantilever (Sampaio & Martins, 2014)	VR-based online module for heavy bridge construction developments based on different steps for bridge engineering for workers.	The understanding of complex issues and steps in bridge construction was performed by the relevant workers.
VR for exploring spaces by blind people (Picinali et al., 2014)	The acoustic cue-based navigation system of a building is developed for blind people in a VR environment along with sensing means.	Spatial mental maps were recorded for the people who provided coherent virtual navigation of buildings easily.

(*Continued*)

Table 5.2 (Continued)

Article focus	*Method and description of the study*	*Outcomes of the study*
Interactive virtual modelling for construction training (Ku & Mahabaleshwarkar, 2011)	4D construction simulation knowledge and collaborative learning are performed to enhance students' understanding of the university.	Multi-user collaboration on developed BIM models provided better education for university students for design purposes.
Multi-user virtual safety training system for tower crane dismantlement (Li et al., 2012)	The dangerous task of crane dismantling is developed virtually in a games-based environment to train workers, citing the fatal nature of this activity and operation.	Different scenarios of crane erection and dismantling were checked based on usability and suitability, and results showed better learning of workers.

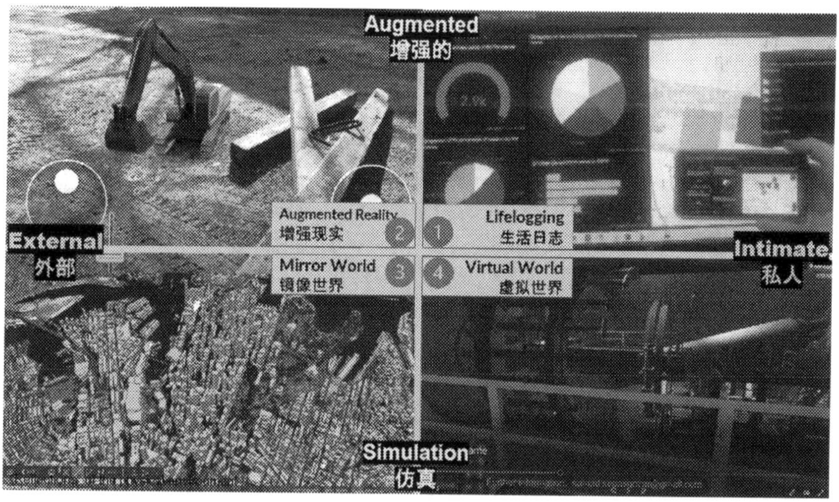

Figure 5.2 Various simulation and augmented technologies for education: (1) lifelogging (see details: Shirowzhan et al. (2021)); (2) augmented reality for construction (source: authors); (3) virtual world (see details: Sepasgozar et al. (2021)); (4) mirror world (source: authors, Cesium).

applications showing different objects or processes in an immersive environment. Examples of the first and third practices are provided in Section 5.3.3 for city planning and city analytic education. Examples of the second and fourth practices for construction and mining education are presented in Sections 5.3.1 and 5.3.2, with similarities noted where heavy equipment and hauling processes are used and in site logistics.

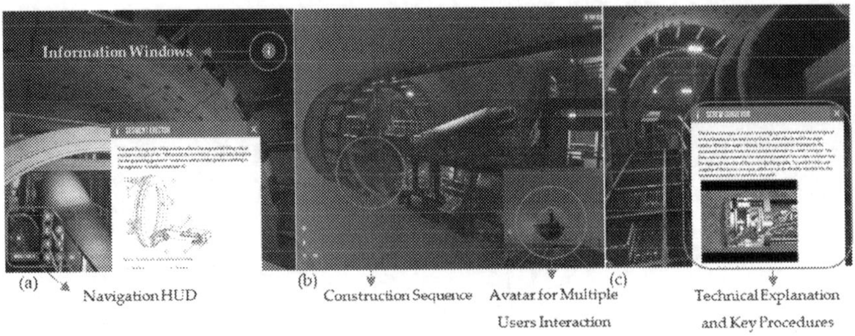

Figure 5.3 The learning function for students. (a) Interactive tools; (b) construction sequence and students' network; (c) technical details of a TBM.

Source: Authors.

5.3.1.1 *Virtual tunnel boring machine*

Other than the highly cited publications published from 2012 to 2018, there are recent advances in the application of immersive technologies and the introduction of DT for education purposes in the very recent literature. For example, Sepasgozar (Sepasgozar, 2020) presented different mixed reality applications for educational purposes in the AEC disciplines. A recent application for training digitally and virtually is integrating gaming technology to develop a pre-DT of a tunnel boring machine (TBM). Figure 5.3 shows details of the environment and some further suggestions that can be used for further improvements of this version of gaming technology.

The virtual simulation work reflects the real machine, providing different features that students can use to learn more about its functionality. Figure 5.3(a) shows how users can navigate and interact with the environment, having a heads-up display (HUD) to quickly move between sections of the TBM and the information windows to learn more about the parts' details. In Figure 5.3(b), the whole excavation and ring installation sequence can be seen, having the advantage of interacting between multiple users at the same time. Finally, in Figure 5.3(c), a detailed explanation of the main parts of the TBM displays how it works as well as what could go wrong if not handled properly.

5.3.1.2 *Pile training module*

A similar approach to the virtual TBM was developed for the pile training module (PTM), implementing it as an educational tool for students and professionals possessing little experience with these procedures. Multiple tools, such as Revit, Unreal Engine, 3Ds Max, AutoCAD, and Sketchup Pro, were used to produce a virtual simulation, as shown in Figure 5.4. Within the PTM, multiple modules showed the required elements and procedures for piling activities on a construction site. Figure 5.4(a) visualises the main construction sequence of piles, showing the important factors and providing tips for selecting drill attachments and weather

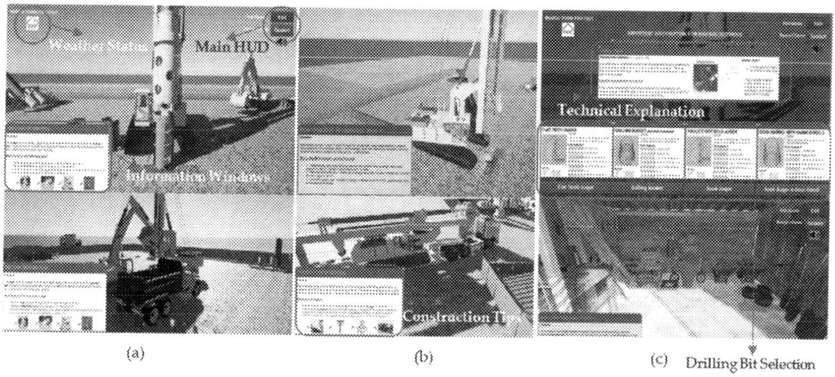

Figure 5.4 The pile training module (PTM). (a) Drilling procedures; (b) reinforcement cage and logistics; (c) drill attachment selection and warehouse elements.

Source: Author.

conditions. In Figure 5.4(b), additional sequences are shown, giving insight into the key factors regarding the logistics and reinforcement erection of pile foundations. Figure 5.4(c) presents the selection HUD for the drilling bits, implementing the technical information contained in windows to assist users in choosing the best accessory and method for optimal excavation under varying weather and soil conditions. Finally, the warehouse of equipment and materials required for pile construction is presented to users, explaining the functionality and importance of each element in pile construction.

In addition, students and professionals were able to select the type of soil and weather for the drilling excavation. These factors affected the selection of drilling bits, each scenario having different behaviours and thus changing the conditions of the optimum alternative for selection. This required the participant to carefully assess the situation in the drilling bits performance window and the soil characteristics, selecting the casing method and the procedure based on their criteria. The outcomes of each stage were measured with a scoring system and a decision tree to give feedback on the selection, explaining why the selected choice was the best one or why it was not adequate for that case.

5.3.1.3 Digital twin excavator and 360° interactive media

Students were able to interact with a 1:100 scale excavator, reviewing the equipment's mechanisms and controls in close detail with a remote controller and an AR simulation. As presented in Figure 5.5, participants had the option to see the machine's hydraulic systems and range of movement in real-time, familiarising themselves with the real equipment's characteristics.

Another approach for active learning is developing interactive 360° images to experience a construction site. The flexibility to move around the inserted images and links will help students learn about the equipment and process details. Figure 5.6 shows an example of these interactive resources created for construction students.

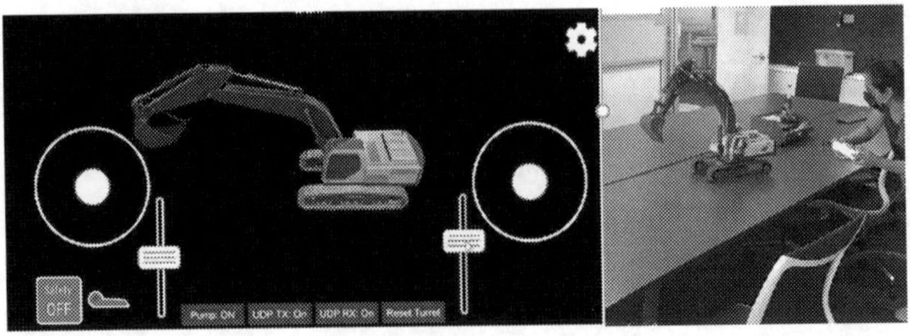

Figure 5.5 Digital twin of an excavator used by students for training purposes. On the right is the physical unit, and on the left is the virtual model of the excavator simulating movements.

Source: Authors' application.

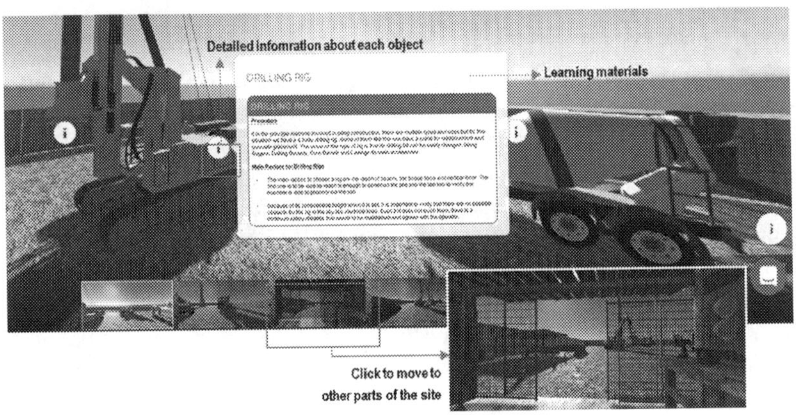

Figure 5.6 Piling virtual tour (PVT) via interactive 360° images.

Source: Author.

5.3.2 *Mining industry*

The construction and mining industries share several similar processes, including excavation, drilling, hauling, and boring, which involve several 3D objects and behaviours that can enhance student learning. The study developed a prototype system for intelligent control of remote mining machines. It is crucial for VR-based education development since it incorporates the logical flow of human-computer interaction.

European universities have carried out major strides in developing VR education tools for mining engineering. The EU-funded MiReBooks project is examining how mining is currently taught, as well as estimating its potential for future changes (Daling et al., 2020; Thurner et al., 2021). By incorporating AR and VR experiences into traditional paper-based teaching materials, educators can now present phenomena in the classroom or lecture hall that would otherwise be difficult to access in the real world, giving students a stimulating opportunity to learn and gain a deeper understanding of mining concepts.

Other studies have developed two VR systems for training and learning (Janiszewski et al., 2020; Janiszewski et al., 2021). The first, virtual underground tunnel environment (VUTE), is a VR learning system based on a photorealistic 3D model of the Underground Research Laboratory at Aalto University and virtual replicas of mapping tools (Janiszewski et al., 2020). The second system, Mining Education and Virtual Underground Rock Laboratory (MIEDU), is designed to train students in structural mapping of discontinuities and rock mass characterisation. It contains two modules: the first teaches discontinuity planes on the tunnel surface, while the second, QVR, focuses on teaching rock characterisation using the Q-system developed by other authors for classifying rock masses in underground openings and field mapping (Barton, 2002; Nick Barton, 1988).

Another VR education system has been developed to help students identify rocks and minerals. This system aims to turn Aalto's rock and mineral sample collection into a digital online learning resource using SfM photogrammetry. After conducting a virtual learning feasibility test, the team at Aalto University found that exercises were completed 50% faster and with a significant reduction in the scatter of answers compared to a control group without VR training.

Grabowski and Jankowski (2015) implemented VR training for 21 mining industry workers to address occupational health and safety concerns in the mining industry. The experiment was successful in convincing training facility owners to implement VR training for younger miners, as they experienced the highest number of workplace accidents. Additionally, the study aimed to provide a comprehensive evaluation of VR training applications.

Research led by some researchers (Abdelrazeq et al., 2019; Kalkofen et al., 2020) created a virtual mine (VR-mine), designed to simulate the learning content of mining progress. VR-mine comprises three educational modules: (1) introduction to mining, (2) fundamentals of mine safety, and (3) rock breaking methods. The system utilises advanced head-mounted displays to offer immersive visualisation of 3D models, allowing students to explore and interact with the virtual environment as if they were at a real mine site. This innovative approach provides students with the opportunity to learn about the challenges and complexities of mining in a safe and controlled environment. Moreover, VR-mine offers a unique experience by allowing students to witness events, such as explosions, that would be too dangerous to demonstrate in a real-life setting.

Liang et al. (2019a) conducted another study in which they developed a VR-based "Serious Game" designed to address rock-related hazards in the mining industry. By incorporating two modules, one for scaling training for novice miners and the other for rock-related hazard perception training for experienced miners, they were able to transfer safety knowledge and enhance interactive safety training effectively. The training has the potential to significantly improve safety conditions in underground mines and evaluate the miners' level of safety awareness and risk aversion in the future (Liang et al., 2019b).

In a more recent study, Beloglazov et al. (2020) introduced the concept of DT and demonstrated how it can be applied in the mining industry to improve the control, prediction, and safety of production objects. The study also emphasised the importance of preserving automation sustainability to reduce technological deviations and accidents at large mining and processing plants. The authors highlighted that integrating virtual and extended reality with DTs can enhance the visualisation of project solutions and hidden process flow elements, thereby improving personnel training. This approach enhances labour safety by providing a visual presentation and enabling trainees and operators to interact with real-life process flows or production objects remotely from training rooms.

VR-based mining education is becoming increasingly important in China's teaching and training about mining (Hou et al., 2021; Liang et al., 2018; Liang et al., 2019a; Yang et al., 2019). Northeastern University (NEU) researchers have developed a VR-based mining education workflow. NEU's team has built a virtual simulation system for underground mining using a precise 3D model to simulate both underground and open-pit mining methods. The system has received positive feedback from students who have become more interested in VR teaching and who have a better understanding of mining methods due to the immersive scenarios. Scholars at the China University of Mining and Technology (CUMT), Beijing, have also created a VR roaming system for coal mining that allows for teaching scenarios outside of actual mine sites (Hou et al., 2019; Zhang et al., 2020). This system is designed to make traditional roadway identification, mining device recognition, and production operation roaming safer and more effective.

Gürer (2021) proposed an example of using game-oriented teaching for underground mining occupational health and safety training. The underground coal mine simulation enables users to navigate for educational purposes, testing their ability to identify existing risks and their knowledge of critical issues through interactive tools within the game. As mines are high-risk workplaces, VR-game training has replaced traditional training methods in some areas. The virtual training environment does not involve real-life risks, and production can continue without interruption. The training allows every miner to be taught the same scenario, reducing its cost and ensuring standardisation of the training provided to all employees, which is difficult to achieve in real life.

5.3.2.1 Advanced visualisation and interaction environment

Among existing applications for mining education, the Advanced Visualisation and Interaction Environment (AVIE) at the University of New South Wales (UNSW) has recorded a number of achievements. The AVIE system contributes to high-quality education based on various mining and petroleum engineering modules, such as coal burst, petroleum awareness, and laboratory rock testing. Considering the teaching and training circumstances, the system is designed based on multi-player technologies. It could host up to 30 visitors through various devices, like the Oculus series, HoloLens, and mobile devices. It has a tailored training program for customised requirements, which could be developed for anyone without a mining background based on current modules. The devices in AVIE are outstanding and world-class. A range of VR theatre technology provides a suitable immersive experience, including EPICylinder Immersive VR Theatre and Projector-based 360° VR Theatres. Currently, there are more than 20 modules available in the School of Minerals and Energy Resource Engineering, including Unaided Self-Escape,

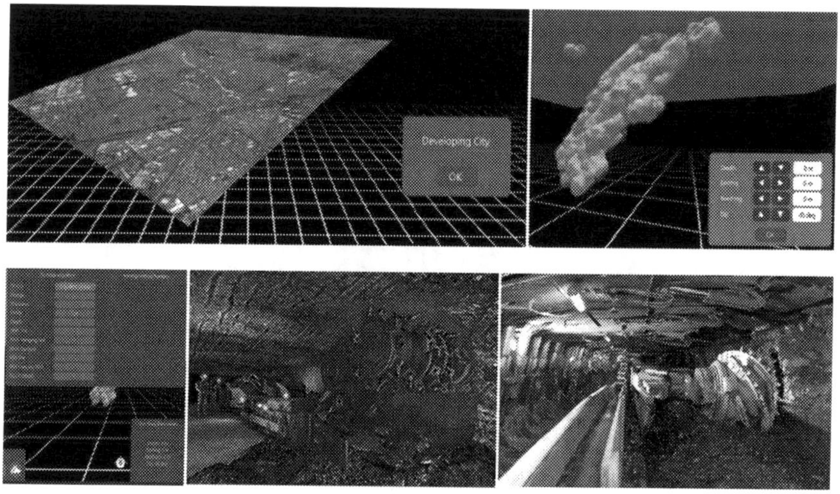

Figure 5.7 The mining method selection system in ViMine (top), examples of what represents mine simulations using a VR look-like (bottom).

Source: Author.

Hazards Awareness, Gas Outburst simulation, 3D presentation of Rib and Roof Stability, Laboratory Rock Testing Procedure Simulation, mining method simulation (block caving mining system visualiser, longwall top coal caving), and an iterative ViMine module, as shown in Figure 5.7.

One example in AVIE that is used to improve students' learning experience in mining methods and system selection is the ViMine system. It supports activities, such as (1) familiarising themselves with the general context of a project scenario; (2) interpreting and using exploration data provided; (3) following recommended mining method selection processes using Excel tools; (4) choosing the best possible methods; (5) using the visualiser function in ViMINE to see how each of these methods would work with the ore deposit; and (6) taking into account other economic and environmental factors identified in the project scenario.

5.3.2.2 Mines of the future

Apart from the traditional teaching areas, VR/AR also provides an opportunity to immerse users into a scene of what the "Mines of the Future" could look like. The visualisation could be used to interact with examples of future mining technology, visualisation of big data (Figure 5.8), and how mining plays an essential role in the modern energy transition effort. For instance, one of the AR modules focuses on the Lithium Open Pit Mining Method. This module illustrates the process of extracting lithium, a critical component in electric car batteries. By placing 3D models of the mining sequence on a table and using AR technology, students can explore the steps involved in the open-pit mining process. The immersive experience enables students to examine the mining operations from various angles, enhancing their understanding of the subject matter.

Figure 5.8 Example images of the use of an immersive VR theatre for Big Data visualisation. The AR module shows the open pit model on a table for lithium extraction.

Source: Author.

Those VR/AR applications have significantly improved students' learning experience, particularly during online delivery. For example, some general feedback from students is: *Extremely enthusiastic and encouraged as much interaction online as possible. The first lecture we had was face-to-face and was excellent, with the incorporation of discussions and AR presentations, as well as real-world examples of the different methods and innovations. A visit to the virtual reality room to get a feel of how coal/rock bursts manifest was a great thing.*

5.3.2.3 DTIT and mining

The systematic search shows 13 publications indexed in Scopus focused on DTIT integration in mining industry education and training. Table 5.3 summarises the description, methodology, and outcomes of these selected publications. Mining is considered a hazardous and risky sector; thus, training and education through DTIT applications become important early in the overall process to avoid unnecessary accidents and even calamities (Beloglazov et al., 2020).

As the number of identified publications in the mining sector was not enough to provide a detailed synthesis of the study, articles from Google Scholar were also considered. The outcomes in terms of DTIT applications in different stages and methodology applied by the researchers are presented in Table 5.4, covering the publications from Scopus and Google Scholar in the mining sector.

5.3.3 Urban planning

This section presents the current studies in the field of urban studies through these main keywords: urban planning, city planning and city analytics. The application of DDIT has received high demand in recent years. There were five identified

Table 5.3 Description of mining publications from Scopus

Article focus	*Method and description of the study*	*The outcome of the study*
DT concept for operator training (Beloglazov et al., 2020)	Explore the potential of DT in improving the control, prediction, and safety of production objects and their operations while also addressing technological deviations and accidents.	The goal is to combine VR technology with DT to create visual representations of project solutions and hidden aspects of process flows. This will allow for a remote, virtual presentation of real-life process flows or production objects while in training rooms.
Reduce performance variability in a VR mining simulator (Nickel et al., 2019)	This study aimed to compare the difference in performance variability among users in the mining industry before using a mining simulator by conducting tutorial sessions. For this purpose, eight participants were randomly divided into a tutorial group and a non-tutorial group.	The tutorial group consistently identified collisions and perceived response times, making it a promising indicator for enhancing worker training in this area. Over time, their variability decreased, suggesting that they were improving in these skills.
Safety training for underground rock-related hazards (Liang et al., 2019b)	The Serious Game, developed for VR, aims to improve safety knowledge and provide interactive training for rock-related hazards. The game features two modules: one for novice scaling training and the other for rock-related hazard perception training for experienced miners.	VR training has emerged as a more effective alternative to instructional videos in improving the safety situation of underground mines. By providing a more immersive and interactive learning experience, VR training has the potential to enhance the evaluation of safety awareness and risk aversion among miners.
VR framework for mining research (Bellanca et al., 2019)	The VR-Mine framework was developed to expedite the creation of an underground mine to collect, simulate, visualise, and train human data. The framework includes studies on mine generation, simulated networks, proximity detection systems, real-time ventilation models, self-escape, and proximity detection.	VR-Mine facilitates the growth and enhancement of mining aspects, its synchronisation with advanced real-time atmospheric monitoring systems (AMS), virtual lighting assessment, the design of a digital replica, and the evaluation of haul truck operators' conduct.

(*Continued*)

Table 5.3 (Continued)

Article focus	*Method and description of the study*	*The outcome of the study*
VR safety training framework (Van Wyk & De Villiers, 2019)	Suggest a framework for assessing the design, creation, and execution of interactive VR training systems as a groundbreaking method to enhance safety training, particularly in the areas of usability and instructional design within the mining sector.	The use of the resulting framework as an assessment tool has improved safety training at 15 training centres across various mines and smelting plants throughout South Africa.
VR system for coal-mining operations (Xie et al., 2018a)	Offer a VR remote operation system for a fully automated coal-mining face, leveraging real-time data and collaborative network technology.	The risk of underground accidents can be reduced by improving miners' training levels and experience.
VR training for the mining industry (Zhang, 2017)	Evaluates the impact of screen-based and VR-based training methods on safety and production in the mining industry by engaging ten trainees in each of the two scenarios.	The outcome demonstrates that the VR-based system provides a more user-friendly and effective experience for users, which is straightforward and enhances overall mine safety.
VR system for accident escapes in mining (Tan et al., 2015)	A VR-based safety training system designed for coal and gas outburst accidents among mining workers.	The system effectively demonstrates impressive escape routes, instructs miners on how to assist themselves, and provides a deep understanding of coal and gas outburst accidents, ultimately minimising the severe consequences that may arise from such incidents.
Study on VR, artificial intelligence and fuzzy logic in the mining industry (Mitra & Saydam, 2014)	Examines the combination of artificial intelligence and fuzzy logic with VR in the mining sector, focusing on the various ways these technologies can be integrated.	The findings reveal the potential ways in which AI and fuzzy logic can be integrated with VR in the mining industry to enhance both learning and teaching outcomes.

(*Continued*)

Table 5.3 (Continued)

Article focus	*Method and description of the study*	*The outcome of the study*
VR-based rescue drill system in coal mines (Lei et al., 2014)	The goal is to create a VR emergency drill system to improve the tactical abilities of rescue teams in the coal-mining industry, specifically in situations where there has been a gas explosion and fire.	The China Kailuan Group's emergency rescue base has implemented the developed system, which has demonstrated its effectiveness in enhancing the rescue capabilities of the participants while ensuring the safety of the training procedures.
Multi-agent-based VR system for coal mines (Tang et al., 2014)	Introduces the concept of safety in mining production through the application of VR and multi-agent technology, utilising a three-tiered architecture model consisting of data, multi-agent, and human-machine interface layers.	The notion encompasses the intricate behaviours of virtual miners in a virtual coalmine environment, including perception, information processing, learning, behaviour, planning, decision-making, and knowledge base establishment. The relevance of this concept is emphasised by its potential to facilitate safe production in mining areas.
VR training for coal miners (Grabowski & Jankowski, 2015)	Conducted VR training for 21 mining industry employees to address occupational health and safety issues during the mining process.	The experiment offers a comprehensive assessment of VR training on a large scale, and it encourages training facilities to provide fundamental training for young miners due to the high rate of workplace accidents among them.
Impact considerations of an incident in mining based on VR simulation (Webber-Youngman & Van Wyk, 2013)	Using VR simulations to recreate mine incidents allows employees to visualise the potential consequences of being exposed to specific hazards.	According to simulations, incorporating mine-related incidents as part of a risk management strategy can lead to a decrease in the number of incidents and should, therefore, be considered for utilisation in preventing such incidents.

publications from Scopus focused on DTIT integration in urban/city planning and city analytics for education and training purposes. Table 5.5 summarises the studies' description, methodology, and outcomes.

Lopes and Lindström (2012) created a virtual environment of part of Uppsala, Sweden, to improve stakeholders' engagement for a novel personal rapid transit (PRT) system. They used a virtual environment and deployment of the developed

Table 5.4 **Applications in the mining industry and education**

Sr. no.	*References*	*VR*	*AR*	*MR*	*With gaming*	*Application*	*Stage*		*Methodology*				
							D	*C*	*R*	*F*	*S*	*CS*	*I*
Scopus													
1	Beloglazov et al. (2020)	✓			★	Training and visualisation for safety	✓	✓	★	★			
2	Liang et al. (2019b)	✓			★	Education and training for rock-related hazards		✓				★	★
3	Nickel et al. (2019)	✓			★	Education for mining simulator training	✓				★	★	
4	Bellanca et al. (2019)	✓			★	Development of VR-mine framework for training		✓		★		★	
5	Van Wyk and De Villiers (2019)	✓				Evaluation framework for safety training		✓		★		★	
6	Xie et al. (2018a)	✓				Mechanised framework for coal mine operations		✓		★		★	
7	Zhang (2017)	✓			★	Intuitive VR training for mining operations		✓			★	★	★
8	Tan et al. (2015)	✓			★	Training for coal and gas-based outburst accidents	✓	✓		★		★	
9	Lei et al. (2014)	✓			★	Rescue drill training of the workers in coal mine fields		✓			★	★	★
10	Grabowski and Jankowski (2015)	✓			★	Pilot training for underground coal miners		✓			★	★	

11	Mitra and Saydam (2014)	✓				Integrating AI and fuzzy logic with mine training	✓	✓	★			
12	Tang et al. (2014)	✓				Framework for mine safety through multi-agent technology	✓	✓	★	★		
13	Webber-Youngman and Van Wyk (2013)	✓			★	Reducing future accidents by simulating impacts		✓			★	★
Google Scholar												
1	Pedram et al. (2021)	✓			★	Immersive training for mine rescuers		✓			★	★
2	Li et al. (2020)	✓			★	Safety training in coal mines integrating AI and cloud computing	✓	✓		★		
3	Andersen et al. (2020)	✓			★	Simulator for mining evacuation drills		✓		★		
4	Daling et al. (2020)	✓	✓	✓		Integration of VR/AR/MR in mining education	✓	✓	★	★	★	
5	Stothard et al. (2019)			✓		Mixed reality potentials for the mining industry			★	★		
6	Hui Zhang et al. (2019)	✓				Fuzzy evaluation of mine safety training system				★		
7	Abdelrazeq et al. (2019)	✓				VR educational tool for the mining industry		✓	★	★		

(*Continued*)

Table 5.4 (Continued)

Sr. no.	*References*	*VR*	*AR*	*MR*	*With gaming*	*Application*	*Stage*		*Methodology*				
							D	*C*	*R*	*F*	*S*	*CS*	*I*
8	Xie et al. (2018b)	✓			★	Framework for VR enabled coal mine simulator				★			
9	Azaryan and Azaryan (2019)		✓			Opportunities for AR integration in mining	✓	✓	★			★	
10	Shiva Pedram et al. (2016)	✓				Evaluation of VR as a safety tool in the mining industry	✓	✓	★	★			
11	Stothard and Laurence (2014)	✓			★	Mining industry operations in large screen VR systems		✓		★	★		
12	van Wyk and de Villiers (2014)	✓			★	Framework for VR training in the mining industry	✓	✓		★		★	
13	Bassan (2011)		✓		★	AR application in mining industry education	✓	✓	★			★	
14	Xiaoqiang et al. (2011)	✓			★	Application framework through VR in the mining face		✓		★		★	
15	Varadarajan (2011)	✓	✓		★	Controlling robots through VR and AR in mining	✓	✓		★		★	
16	Bednarz et al. (2011)	✓	✓			VR for teleoperation and tele-assistance in mining		✓		★			
17	Ennifer Tichon (2011)	✓				Review of VR as a safety training tool in mining	✓	✓	★				

18	Etienne van Wyk (2009)	✓	★	VR training applications' study in the mining sector		✓	★			★	
19	Keping Zhou (2010)	✓	★	Simulator system for underground mining	✓	✓		★		★	
20	Orr et al. (2009)	✓	★	Fire escape training in mining operations with VR		✓				★	
21	Kizil (2003)	✓		VR applications in the Australian mining sector			★				
22	Kizil et al. (2004)	✓	★	Training and education of mining operations with VR		✓	★	★		★	
23	Squelch (2001)	✓	★	Safety in mines using VR training in South Africa	✓	✓		★		★	★
24	Bennett et al. (2010)	✓		VR effectiveness in learning mining operations		✓				★	
25	Mallett and Unger (2010)	✓		Review of VR training in the mining industry		✓	★				

Note: D – design, C – construction, R – review, F – framework, S – survey, CS – case study, I – interview.

PRT system to obtain its viability as an urban transportation system. The results of the case study of the use of the PRT system, including car pods in Uppsala City, received complimentary feedback from the stakeholders, using them to contribute towards the formation of a smart urban Uppsala City.

A study by Monther and Jamhawi (2016) utilised ICT, VR, and GIS to develop an app that guides tourists through the historic sites in Madaba, Jordan. Two workshops were held to gather feedback and measure the app's usability. The virtual tour received positive feedback and was deemed effective in attracting more tourists to the city. Tunçer (2020) recent study used augmented reality to examine the impact of recreational spaces and road categorisation on walking in Singapore. Social media data was analysed using a machine learning algorithm to select those related to art. The study also explored the importance of human interactions in cities, which are key factors in urban livability, sustainability, and economic growth. AR was found to be highly effective in design experiments where sensory experiences play a role in the participants' experience and can be integrated into urban planning and design processes (Redondo et al., 2014). Table 5.5 provides a summary of selected publications.

In addition to this, some researchers (Villena Taranilla et al., 2019) examined the eventual benefits of a selected DTIT called Vir Time Place in the context of urban history focusing on the Roman Empire. They reported that there is a statistically significant difference in favour of learning with DTIT in terms of motivation and performance. Meanwhile, DT and MR have received attention recently in the context of urban study education, although recently, 3D modelling and GIS have been used extensively. Minnery and Searle (2014) examined the application of Sim City 4 for learning and reported that it was beneficial to the game players. However, they reported that the considerable simplification of city processes within the Sim City 4 environment was not able to offer students a sense of realistic decision-making or planning.

A likely future direction could be the combination of recent versions of VR and previous 3D modelling and information systems. This will offer 360-degree panoramic VR experiences in an ArcGIS environment. GIS data can be visualised using City Engine or ArcGIS Runtime. City Engine may support ArcGIS 360 VR, City Engine VR experience, the Unity or Unreal game engine, and efficient interoperability, which is very important to users. Çöltekin et al. (2020) reviewed the extended reality technologies in spatial sciences, including AR sandbox. The AR sandbox demonstrated in Figure 5.9 is also being used to teach topography concepts and works in collaborative environments for geo-designing and redevelopment assessment projects. The sandbox has been found to be very interesting for students and professionals engaged in these projects since it leads to better communication of their mapping ideas in the assignments and projects, they are involved in.

Unlike the mining discipline, the previously identified studies in urban/city planning and city analytics did not provide a detailed synthesis of the study. Therefore, articles from Google Scholar were also considered. The outcomes in terms of DTIT applications in different stages and for different methodologies applied by

Table 5.5 Description of urban planning, city planning, and city analytics publications from Scopus

Article focus	*Description and method of the study*	*The outcome of the study*
Urban landscape simulation based on VR and machine learning (Piao et al., 2020)	The conceptual framework was derived by integrating VR and machines for urban landscape planning around a subway station, focusing on living arrangement time calculations.	The outcomes display prompt data to aid in creating practical and accurate designs and offer high-quality smart illustrations.
AR-based big data-informed urban design and planning (Tunçer, 2020)	Utilised AR to identify uncovered recreational areas, natural landscapes, and specific road types as destinations for leisurely walks within the urban street network in Singapore. An algorithm based on machine learning was applied to gather social media feeds and select those that pertained to art.	AR has proven to be highly effective in design experiments where the sounds, sensations, and smells of the context play a crucial role in the participants' experience and are integrated into planning and urban design processes.
IT-based innovation and technology transfer in heritage sites (Jamhawi and Hajahjah, 2016)	The study aims to enhance the tourism industry in Madaba City by using Information and Communication Technologies (ICT), VR, and GIS to create an app guiding tourists to historic sites in the city.	The feedback from workshop attendees regarding the prototype was overwhelmingly positive, with participants citing high user satisfaction and effectiveness levels. Furthermore, the virtual tour component of the prototype was well-received, with many participants indicating that it would be an effective tool for attracting more tourists to visit Madaba.
3D navigation-based user learning of geo-referenced scenes (Yiakoumettis et al., 2014)	Proposes active learning of user's preferences in urban planning for the best route selection process and learning strategies by categorising the weights based on the feedback of metadata provided by the users.	The use of a learning strategy resulted in an average improvement of 13.76% in precision, while the route selection strategy improved by 8.75%. Additionally, the weight rectification strategy helped reduce the number of user samples to achieve higher precision.
VR-based urban planning case study (Lopes & Lindström, 2012)	Aims to get stakeholder engagement for a novel personal rapid transit (PRT) system in Uppsala City, Sweden, by creating a virtual environment of a part of the city and deploying the developed PRT system.	PRT system, including car pods in Uppsala City, received positive feedback from the stakeholders regarding deploying these systems, which will contribute to the smart urban city.

Table 5.6 **Application in urban planning, city planning, and city analytics**

Sr. no.	*References*	*VR*	*AR*	*MR*	*With gaming*	*Application*	*Stage*		*Methodology*				
							D	*C*	*R*	*F*	*S*	*CS*	*I*
Scopus													
1	Lopes and Lindström (2012)	✓			★	VR simulation of parts of a city	✓					★	
2	Yiakoumettis et al. (2014)	✓				3-D visualisation route planning	✓			★		★	
3	Jamhawi and Hajahjah (2016)	✓			★	Virtual tour for tourism planning	✓				★	★	
4	Tunçer (2020)		✓		★	AR-based urban design & planning	✓					★	
5	Piao et al. (2020)	✓				VR simulation of the urban landscape	✓			★			
Google Scholar													
1	Du et al. (2020)	✓				Interactive, immersive public displays in urban planning	✓				★		★
2	Kent et al. (2019)	✓		✓	★	Citizen engagement in urban planning with VR and MR	✓			★	★	★	
3	Nieto-Lugilde (2019)	✓			★	A serious game for urban planning learning and education	✓			★	★		

4	West et al. (2019)	✓			★	Enhancing citizen engagement through VR in smart cities	✓			★	★		
5	LeMoine (2017)	✓	✓		★	Urban planning and redesign engagement using mixed reality techniques	✓	✓	★		★		
6	Fonseca et al. (2017)	✓			★	Student motivation to use VR in urban planning education	✓			★		★	★
7	Fonseca et al. (2017)	✓			★	Game-based learning for urban planning education	✓	✓				★	
8	Nguyen et al. (2016)	✓			★	City planning walkthrough using VR techniques	✓			★		★	
9	Hisham et al. (2015)		✓	✓	★	Visualising urban planning projects with AR and MR techniques	✓			★		★	
10	Redondo et al. (2014)		✓		★	AR assessment as a visualiszing medium in urban planning	✓	✓			★	★	★

(*Continued*)

Table 5.6 (Continued)

Sr. no.	*References*	*VR*	*AR*	*MR*	*With gaming*	*Application*	*Stage*		*Methodology*				
							D	*C*	*R*	*F*	*S*	*CS*	*I*
11	Poplin (2012)	✓			★	Serious gaming for public participation in urban planning	✓			★	★		★
12	Shen (2012)	✓				Application of VR in urban planning and management	✓	✓	★	★			
13	Praveen Maghelal et al. (2011)	✓			★	VR application pedestrian movement in urban planning	✓				★	★	★
14	Shailey Minocha (2010)	✓			★	Design guidelines for usability and navigation of 3D urban spaces	✓			★			★
15	Phan (2010)		✓		★	AR design platform for collaborative work in urban space	✓			★		★	
16	Reaver (2020)		✓	✓	★	AR and MR use evaluation in urban planning projects	✓					★	
17	Redondo et al. (2020)	✓			★	VR-based serious game for student urban design education	✓	✓			★	★	

Note: D – design, C – construction, R – review, F – framework, S – survey, CS – case study, I – interview.

the researchers are presented in Table 5.6, covering the documents from Scopus and Google Scholar in the relevant sector.

5.4 Discussion

This chapter reviewed the DTIT applications for education and training in three selected disciplines. The contribution of this chapter is to identify various simulation, visualisation, or immersive technologies used in three different disciplines with the practical components to be covered. In particular, the chapter presented six DTIT modules developed for the three selected disciplines, including PTM, excavator DT, PVT, ViMine, Big Data visualisation, and AR sandbox. In addition, the chapter suggests directions for future studies based on the systematic review and the developed modules.

Recently, ArcGIS 360 VR or VR applications associated with BIM have been introduced to the market, including real data or objects that should be used for training purposes. The simulation or digital part of the industry DT can be used for active and authentic learning (Broo & Schooling, 2021).

To achieve significant results out of DTIT applications, a theoretical lens or conceptual models are also required to assess the students' learning improvement or users' behaviour and their intention of using the virtual modules. Table 5.7 shows a list of concepts or theories that can be modified to examine students' learning and perceptions while using the virtual modules. While these models are useful for evaluating users' perceptions, they are not designed to evaluate students' acquired knowledge, including their overall usefulness, the DTIT system's level of immersion, usability, or efficiency of the DTIT system, and task performance.

The review shows that current investigations are divided into two major groups in terms of the purpose of experimentation, that is (i) examining learning experience and improvements and (ii) measuring user satisfaction. Many of the current theories in the field of education are used to examine students' learning experience, such as constructive learning theory (Huang et al., 2010; McHaney et al., 2017) and experiential learning theory (Fromm et al., 2021; Huang et al., 2016). Other researchers have borrowed a conceptual model from the information system field and are focused on user satisfaction, covering factors such as usefulness, ease of use, immersion, enjoyment, and dizziness, such as TAM, presence theory (Ke et al., 2016; von der Pütten et al., 2012), situation cognition theory (Chang et al., 2016; Dawley & Dede, 2014), and cognitive load theory (Grant Frederiksen et al., 2020; Hsu, 2017). For example, Kim et al. (2020) reviewed a set of key factors such as head tracking, eye tracking, body tracking, microphones, pressure sensors, and tracking handheld controllers. Kim et al. (2020) also reported that user interaction factors such as the ability to connect with others (multi-user), flexibility to sit or stand, and interaction options affect user satisfaction with a DTIT. Overall, user experience is affected by individual background (e.g., knowledge, gender, age, experience, and personality), technology attributes, and user needs and activity.

Table 5.7 Theoretical models/theories for measuring students'/workers' learning ability and degree of improvement

Theories related to DTIT applications	*Brief description of the theory*	*Key measures*
Cognitive load theory (Grant Frederiksen et al., 2020; Hsu, 2017)	Describes the cognitive loads experienced by the users while using DTITs because of the large data, information, and technological devices used during the process.	Intrinsic cognition – load/effort related to a precise theme; extraneous cognition – load/effort in grasping the information of a specific task, and germane cognition – load/effort required to store the information delivered.
Flow theory (Bian et al., 2016; Chang et al., 2014; Georgiou & Kyza, 2017)	Describes the flow state experienced by the users while using DTITs, which includes high concentration, a sense of positive enjoyment, the balance between challenge and skill, positive experience time, and immersive flow.	Measures to perceive what to do, how to do it, navigation flow, challenges, skills required, and liberty from interruptions and disruptions.
Conceptual blending theory (Enyedy et al., 2015; Gregorcic & Haglund, 2021)	Describes the conceptual blend experienced by the DTIT users, which allows fluid movement between real and synthetic space. Also, learning is enhanced as the user tends to differentiate between sources, distinguish spaces and coordinate across multiple areas.	The key factors for blending include the composition – relation between different elements in a space; completion – additional features and gestures related to various elements; and elaboration – smooth display of virtual simulation in terms of information delivery.
Constructive learning theory (Huang et al., 2010; McHaney et al., 2017)	Describes the user's learning through DTITs as active learning because the user can construct new knowledge in addition to the knowledge and information delivered through an immersive environment.	Measures to achieve constructivism in the output relate to having four factors – concrete experience, observation, and reflection, forming abstract concepts, and testing in new situations.
Experiential learning theory (Fromm et al., 2021; Huang et al., 2016)	Describes enhancing the learning experience of the user in four stages, namely concrete experience, observation and learning, abstract concept formation, and testing new situations.	Key factors include active experimentation, concrete experience, reflective observation, and abstract conceptualisation.

(*Continued*)

Table 5.7 (Continued)

Theories related to DTIT applications	*Brief description of the theory*	*Key measures*
Motivation theory (Di Serio et al., 2013; Tunur et al., 2021)	Investigates users' enhanced stimulus in an immersive environment with increased attention and satisfaction.	Measures related to the user include attention, relevance, confidence, and satisfaction.
Situated cognition theory (Chang et al., 2016; Dawley & Dede, 2014)	Describes the enhanced learning performance experienced by the users through DTITs by practising social and cultural meaning to the context.	Demonstrates and identifies desired results, determines acceptable evidence, and plans learning experiences and instructions.
Technology acceptance model (TAM) (Huang & Liao, 2015; Wojciechowski & Cellary, 2013; Yilmaz, 2016)	Examine the effectiveness of perceived usefulness and ease of technology to the DTITs users.	Factors include perceived usefulness and perceived ease of use for the end-user in an immersive environment.
Presence theory (Ke et al., 2016; von der Pütten et al., 2012)	Describes the perceived sense of presence experienced by DTITs users fostering their motivation and engagement through activities and learning outcomes.	Key measures of presence theory are interaction level, knowledge and experience, affective investment, involvement, and cohesiveness.
Stimuli organism response (SOR) framework (Kourouthanassis et al., 2015; Zhai et al., 2019)	Describes the behavioural changes experienced by DTITs users through cognitive and affective states evoked by technological stimuli.	As the name suggests, the SOR framework's key factors are stimulus – a functional reaction; organism – relating to emotional state and preference; and response – relating to approach or avoidance.

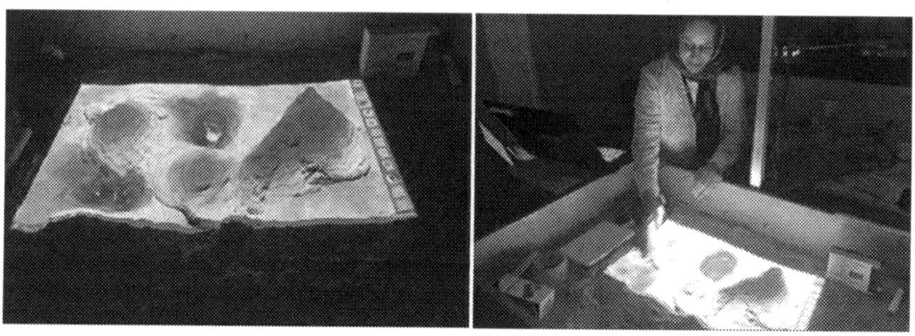

Figure 5.9 AR sandbox for teaching topography mapping and design in the city analytics program.

Source: Author.

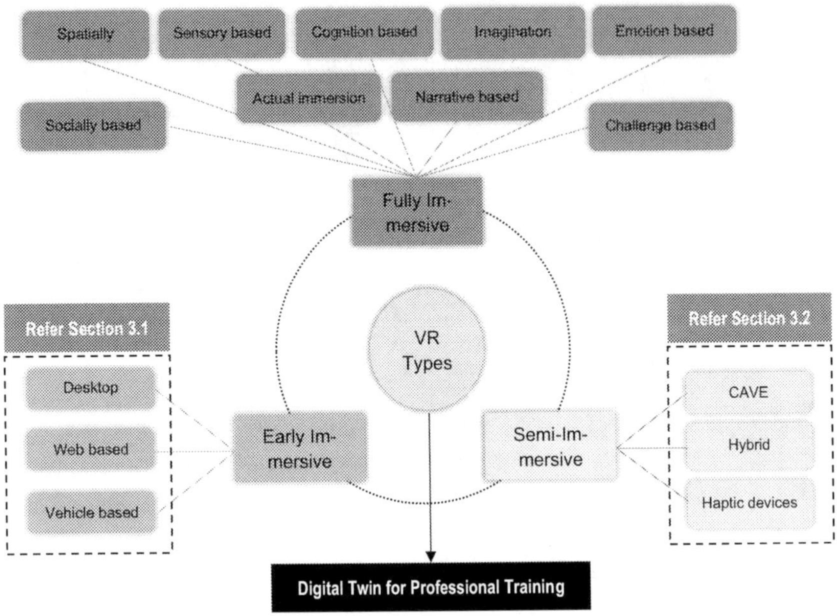

Figure 5.10 Types of VR systems and their immersion basis

Figure 5.10 shows three significant classifications of VR types based on immersion level: early immersive, semi-immersive, and fully immersive modules. Early immersive modules offer user interaction in the form of desktop-based, web-based, and vehicle-based, and examples of these modules are presented in Sections 5.3.1 (in the AEC contexts) and 5.3.3 (in the urban planning context). Desktop-based VR delivers a 3D virtual world on a desktop display without any specific movement or tracking equipment. Elgewely et al. (2021) proposed a desktop VR platform for learning construction detailing in a risk-free environment. Similarly, Beh, Rashidi, Talei, and Lee (2021) developed a game-based desktop VR platform for building utility inspection training for AEC students. On the other hand, web-based VR is a platform, an independent setup through which users can create immersive and interactive applications. Sanchez, Ballinas-Gonzalez, and Rodriguez-Paz (2021) developed a web-based BIM-VR application for the education and learning of engineering students in the AEC domain. Finally, in the early immersive category, vehicle-based VR is significant in the areas where driving simulations of vehicles are required. Although mostly used in the automotive industry, its relevance and use have been seen in the AEC and mining industries, as both these sectors require specific workers to be trained in the context of driving the vehicles in the risk areas on the site (Ahn et al., 2020; Grabowski & Jankowski, 2015).

Semi-immersive modules differ from early immersive modules in that they require the user's movement in a specific environment and the utilisation of tracking devices for gestures. The movement and tracking devices are classified as wireless, optical, inertial, sensor, acoustic, and magnetic tracking, among others (Cipresso et al., 2018). Support devices are head-mounted displays (HMDs) and haptic devices for kinesthetics communication. Examples of these modules are presented in Section 5.3.2 (in the mining context). Finally, fully immersive VR provides the highest level of perception in terms of physical presence. The surrounding continuity, human vision conformation, movement freedom, physical interaction, and feedback are some of the attributes of fully immersive VR. It includes an HMD, a tracking device, and data gloves, among others, as components to generate a 3D animation, giving the user an interactive virtual environment experience (Li et al., 2018). Usually, it also affects sight and sound among the other five human senses in most cases.

Figure 5.11 shows how AR/MR tools can be categorised based on the technique or the development technology, the component involved, the medium worn, and display types. The chapter has provided some examples of wearable and non-wearable modules using various hardware and software programs. The future education modules should be developed in different forms and versions so a wider range of students can use the module in different ways. Table 5.8 shows the current concepts or modules and provides a set of directions for future investigations. While DT is a growing market and is predicted to be widely accepted by practitioners for industry needs, the industry versions can also be used for

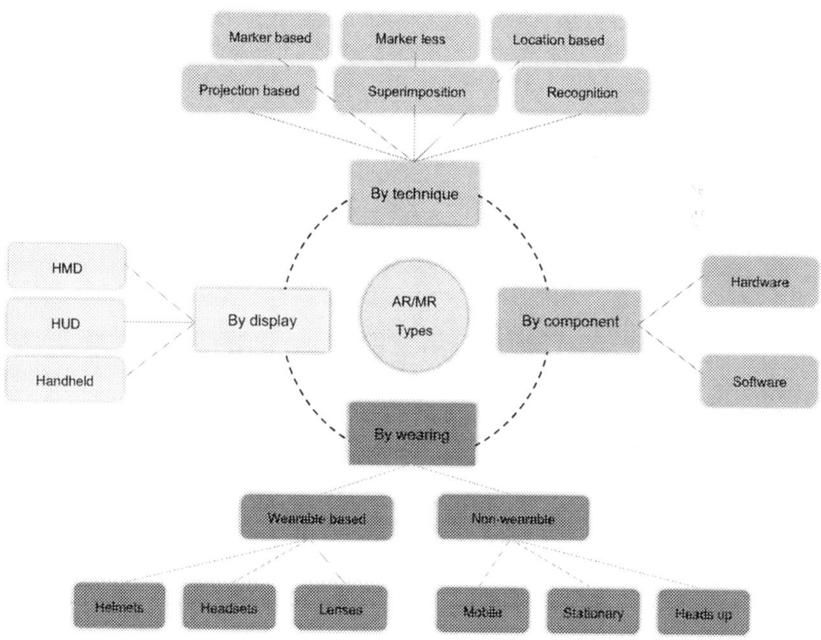

Figure 5.11 Types of AR/MR systems and their immersion basis.

Table 5.8 State of play and future directions

Key factors, state of play, and current publications	*Future directions*
DT allows practitioners to learn from the behaviour of a real entity and use the simulation and virtual models of the DT.Limited studies are focused on the application of DT in education and professional training (Liljaniemi & Paavilainen, 2020; Wang et al., 2020).	Education DT needs to be conceptually and physically developed.It provides firsthand data and supports authentic education approaches due to connections to real practice.This enhances case study-based learning.
A set of theoretical models was used in most educational experimentations (refer to Table 5.7).Recent studies suggest structured acceptance methods to measure users' perceptions of the immersive models (Fromm et al., 2021; Grant Frederiksen et al., 2020; Tunur et al., 2021; Yilmaz, 2016)	Other than measuring learning improvement, users' experience and engagement need to be measured.These need structured predictive models to be examined, including key variables to measure perceived usefulness and ease of use, immersion, attitude, and intention to use the model continuously.
A wide range of evaluations are used to measure users' experience: measuring tools (e.g., qualitative or quantitative), participants (students or experts), experience period (before/after usage, before/after exam), over time, control, and experimental methods.	A robust protocol should be designed and utilised for measuring students' experience.
Ease of use.Reliability and cost are reported as reasons for using some tools, i.e., Oculus Rift and Unity 3D.	Interoperability or compatibility issues of different immersive tools should be resolved.
DT predicts processes and changes in an asset along with scenario analysis (Gomerova et al., 2021; Liljaniemi & Paavilainen, 2020).	A wide variety of individual products, services, and platforms can share knowledge with students, but the collaborative constellation of technologies should have a complete and comprehensive DT for learning purposes.

education. This will address the need for construction site visits, and students can experience the simulated or digital versions of a real DT for their learning purposes.

5.5 Conclusions

Digital technologies are increasingly used in construction, mining, and urban development projects. These technologies include reading datasets from different projects and simulating or representing real changes in various projects, typically used for project or client needs. However, digitally simulated models or the digital representative of projects can also be used for learning purposes. This chapter has reviewed the current literature systematically and presented a set of modules and teaching practices using immersive technologies, extended realities (XR), and DT for learning purposes. The systematic review shows that DTIT modules are widely developed and reported in the AEC literature by 87 publications, which is more than mining and urban planning by 13 and five publications, respectively. The current literature shows that fully or semi-immersive modules are used in the construction and mining disciplines, and early immersive technologies are used in urban planning and related disciplines.

This study is limited to the literature and selected practices of using DTIT for education and presents useful examples. However, the maturity of the DTIT tools and the implementation process of these technologies should be examined scientifically before application in university subjects. Two main constructs that are suggested by technology acceptance theories are usefulness and ease of use. The application of DTIT and these constructs should be considered during curriculum development since they may affect the student's learning experience. The limitations noted do not impair the value of this chapter, and it provides direction for continuing to examine them or advance the DTIT module creations. Since digital technologies are growing, future work needs to seek how to modify the curriculum to embrace advanced technologies.

References

Abdelrazeq, A., Daling, L., Suppes, R., Feldmann, Y., & Hees, F. (2019). *A Virtual Reality Educational Tool in the Context of Mining Engineering-The Virtual Reality Mine.* Paper presented at the Conference Paper. https://www. researchgate.net/publication/332282684.

Ahn, S., Kim, T., Park, Y. J., & Kim, J. M. (2020). Improving effectiveness of safety training at construction worksite using 3D bim simulation. *Advances in Civil Engineering*. https://doi.org/10.1155/2020/2473138.

Alizadehsalehi, S., Hadavi, A., & Huang, J. (2021). Assessment of AEC students' performance using BIM-into-VR. *Journal of Applied Sciences, 11*, 23. https://doi.org/10.3390/app11073225.

Alizadehsalehi, S., Hadavi, A., & Huang, J. C. (2020). From BIM to extended reality in AEC industry. *Automation in Construction, 116*, 103254. https://doi.org/10.1016/j.autcon.2020.103254.

Almeida, F., & Simoes, J. (2019). The role of serious games, gamification and Industry 4.0 tools in the Education 4.0 paradigm. *Contemporary Educational Technology, 10*(2). https://doi.org/10.30935/cet.554469.

Andersen, K., Gaab, S. J., Sattarvand, J., & Harris, F. C. (2020). METS VR: Mining evacuation training simulator in virtual reality for underground mines. In *17th International Conference on Information Technology–New Generations (ITNG 2020)* (pp. 325–332).

Azaryan, A.A. and Azaryan, V.A., 2018. Investigation of Opportunities of the Practical Application of the Augmented Reality Technologies in the Information and Educative Environment for Mining Engineers Training in the Higher Education Establishment.

Barton, N. (1988). Rock mass classification and tunnel reinforcement selection using the *Q*-system. *Rock Classification Systems for Engineering Purposes*. https://doi.org/10.1520/STP48464S.

Barton, N. (2002). Some new Q-value correlations to assist in site characterisation and tunnel design. *International Journal of Rock Mechanics and Mining Sciences, 39*(2), 185–216. https://doi.org/10.1016/S1365-1609(02)00011-4.

Bassan, J. (2011). *The Augmented Mine Worker – Applications of Augmented Reality in Mining*. AusIMM conferences.

Beh, H. J., Rashidi, A., Talei, A., & Lee, Y. S. (2021). Developing engineering students' capabilities through game-based virtual reality technology for building utility inspection. *Engineering, Construction and Architectural Management* (ahead-of-print). https://doi.org/10.1108/ECAM-02–2021–0174.

Bellanca, J. L., Orr, T. J., Helfrich, W. J., Macdonald, B., Navoyski, J., & Demich, B. (2019). Developing a virtual reality environment for mining research. *Mining, Metallurgy and Exploration*. https://doi.org/10.1007/s42461-018-0046-2.

Bednarz, T., James, C., Caris, C., Haustein, K., Adcock, M. and Gunn, C. (2011, December). Applications of networked virtual reality for tele-operation and tele-assistance systems in the mining industry. In *Proceedings of the 10th International Conference on Virtual Reality Continuum and Its Applications in Industry* (pp. 459–462).

Beloglazov, I. I., Petrov, P. A., & Bazhin, V. Y. (2020). The concept of digital twins for tech operator training simulator design for mining and processing industry. *Eurasian Mining*. https://doi.org/10.17580/em.2020.02.12.

Bennett, L., Stothard, P. and Kehoe, J. (2010). *Evaluating the Effectiveness of Virtual Reality Learning in a Mining Context*. SimTect 2010.

Bian, Y., Yang, C., Gao, F., Li, H., Zhou, S., Li, H., ... Meng, X. (2016). A framework for physiological indicators of flow in VR games: Construction and preliminary evaluation. *Personal and Ubiquitous Computing, 20*(5), 821–832. https://doi.org/10.1007/s00779-016-0953-5.

Bouska, R., & Heralova, R. S. (2019). *Implementation of Virtual Reality in BIM Education*. Paper presented at the Advances and Trends in Engineering Sciences and Technologies III - Proceedings of the 3rd International Conference on Engineering Sciences and Technologies, ESaT 2018.

Broo, D. G., & Schooling, J. (2021). Digital twins in infrastructure: Definitions, current practices, challenges and strategies. *International Journal of Construction Management*, 1–10. https://doi.org/10.1080/15623599.2021.1966980.

Chacón, R. (2021). Designing Construction 4.0 activities for AEC classrooms. *Buildings, 11*(11). https://doi.org/10.3390/buildings11110511.

Chang, K.-E., Chang, C.-T., Hou, H.-T., Sung, Y.-T., Chao, H.-L., & Lee, C.-M. (2014). Development and behavioral pattern analysis of a mobile guide system with augmented reality for painting appreciation instruction in an art museum. *Computers & Education, 71*, 185–197. https://doi.org/10.1016/j.compedu.2013.09.022.

Chang, H.-Y., Hsu, Y.-S., & Wu, H.-K. (2016). A comparison study of augmented reality versus interactive simulation technology to support student learning of a socio-scientific issue. *Interactive Learning Environments, 24*(6), 1148–1161. https://doi.org/10.1080/10494820.2014.961486.

Chen, F.-Q., Leng, Y.-F., Ge, J.-F., Wang, D.-W., Li, C., Chen, B., & Sun, Z.-L. (2020). Effectiveness of virtual reality in nursing education: meta-analysis. *Journal of medical Internet research, 22*(9), e18290.

Cipresso, P., Giglioli, I. A. C., Raya, M. A., & Riva, G. (2018). The past, present, and future of virtual and augmented reality research: A network and cluster analysis of the literature. *Frontiers in Psychology, 9*, 2086. Retrieved from https://www.frontiersin.org/article/10.3389/fpsyg.2018.02086.

Çöltekin, A., Lochhead, I., Madden, M., Christophe, S., Devaux, A., Pettit, C., … Hedley, N. (2020). Extended reality in spatial sciences: A review of research challenges and future directions. *ISPRS International Journal of Geo-Information, 9*(7). https://doi.org/10.3390/ijgi9070439.

Dai, C. P., Ke, F., Dai, Z., & Pachman, M. (2023). Improving teaching practices via virtual reality - Supported simulation - Based learning: Scenario design and the duration of implementation. *British Journal of Educational Technology, 54*(4), 836–856.

Daling, L., Kommetter, C., Abdelrazeq, A., Ebner, M., & Ebner, M. (2020). Mixed reality books: Applying augmented and virtual reality in mining engineering education. In *Augmented Reality in Education,* Lea Daling, Christopher Kommetter, Anas Abdelrazeq, Markus Ebner and Martin Ebner. Springer: 185–195. https://link.springer.com/chapter/10.1007/978-3-030-42156-4_10

David, J., Lobov, A. and Lanz, M., 2018, October. Learning experiences involving digital twins. In *IECON 2018-44th annual conference of the IEEE industrial electronics Society.* IEEE: 3681–3686. https://ieeexplore.ieee.org/xpl/conhome/8560606/proceeding.

Dawley, L., & Dede, C. (2014). *Situated Learning in Virtual Worlds and Immersive Simulations*, Springer: 723–734. https://link.springer.com/book/10.1007/978-1-4614-3185-5.

Dembski, F., Wössner, U., Letzgus, M., Ruddat, M., & Yamu, C. (2020). Urban digital twins for smart cities and citizens: The case study of Herrenberg, Germany. *Sustainability, 12*, 17p. https://doi.org/10.3390/su12062307.

Di Serio, Á., Ibáñez, M. B., & Kloos, C. D. (2013). Impact of an augmented reality system on students' motivation for a visual art course. *Computers & Education, 68*, 586–596. https://doi.org/10.1016/j.compedu.2012.03.002.

Du, G., Kray, C., & Degbelo, A. (2019). Interactive immersive public displays as facilitators for deeper participation in urban planning. *International Journal of Human-Computer Interaction, 36*. https://doi.org/10.1080/10447318.2019.1606476.

Du, G., Kray, C., & Degbelo, A. (2020). Interactive immersive public displays as facilitators for deeper participation in urban planning. *International Journal of Human–Computer Interaction, 36*(1), 67–81. https://doi.org/10.1080/10447318.2019.1606476.

Eiris, R., Moore, H., Gheisari, M., & Esmaeili, B. (2018). *Development and Usability Testing of a Panoramic Augmented Reality Environment for Fall Hazard Safety Training*. In *Advances in Informatics and Computing in Civil and Construction Engineering: Proceedings of the 35th CIB W78 2018 Conference: IT in Design, Construction, and Management* (pp. 271–279). Springer International Publishing

Elgewely, M. H., Nadim, W., ElKassed, A., Yehiah, M., Talaat, M. A., & Abdennadher, S. (2021). Immersive construction detailing education: Building information modeling (BIM)–Based virtual reality (VR). *Open House International* (ahead-of-print). https://doi.org/10.1108/OHI-02-2021-0032.

Ennifer Tichon, R. B.-L. (2011). A review of virtual reality as a medium for safety related training in mining. *Journal of Health & Safety Research & Practice*, *3*(1), 33–40.

Enyedy, N., Danish, J. A., & DeLiema, D. (2015). Constructing liminal blends in a collaborative augmented-reality learning environment. *International Journal of Computer-Supported Collaborative Learning, 10*(1), 7–34. https://doi.org/10.1007/s11412-015-9207-1.

Fonseca, D., Falip, S., Navarro, I., Redondo, E., Valls, F., Llorca-Bofí, J., … Calvo, X. (2017). Student Motivation Assessment Using and Learning Virtual and Gamified Urban

Environments. In *Proceedings of the 5th international conference on technological ecosystems for enhancing multiculturality*, Association for Computing Machinery: 1–7.

Fonseca, D., Villagrasa, S., Navarro, I., Redondo, E., Valls, F., & Sánchez, A. (2017). Urban gamification in architecture education. In *Recent Advances in Information Systems and Technologies*. Rocha, Á., Correia, A., Adeli, H., Reis, L., Costanzo, S., WorldCIST 2017. Advances in Intelligent Systems and Computing, Springer: 335–341. https://doi.org/10.1007/978-3-319-56541-5_34.

Fromm, J., Radianti, J., Wehking, C., Stieglitz, S., Majchrzak, T. A., & vom Brocke, J. (2021). More than experience? - On the unique opportunities of virtual reality to afford a holistic experiential learning cycle. *The Internet and Higher Education, 50*, 100804. https://doi.org/10.1016/j.iheduc.2021.100804.

Georgiou, Y., & Kyza, E. A. (2017). The development and validation of the ARI questionnaire: An instrument for measuring immersion in location-based augmented reality settings. *International journal of human-computer studies, 98*, 24–37. https://doi.org/10.1016/j.ijhcs.2016.09.014.

Gomerova, A., Volkov, A., Muratchaev, S., Lukmanova, O., & Afonin, I. (2021). *Digital Twins for Students: Approaches, Advantages and Novelty*. Paper presented at the 2021 IEEE Conference of Russian Young Researchers in Electrical and Electronic Engineering (ElConRus).

González-Zamar, M.-D., & Abad-Segura, E. (2020). Implications of virtual reality in arts education: Research analysis in the context of higher education. *Education Sciences, 10*(9), 225.

Grabowski, A., & Jankowski, J. (2015). Virtual reality-based pilot training for underground coal miners. *Safety Science*. https://doi.org/10.1016/j.ssci.2014.09.017.

Grant Frederiksen, J., Dreier Sørensen, S. M., Konge, L., Svendsen, M., Nobel-Jørgensen, M., Bjerrum, F., & Andersen, S. (2020). Cognitive load and performance in immersive virtual reality versus conventional virtual reality simulation training of laparoscopic surgery: A randomized trial. *Surgical Endoscopy, 34*, 1–9. https://doi.org/10.1007/s00464-019-06887-8.

Gregorcic, B., & Haglund, J. (2021). Conceptual blending as an interpretive lens for student engagement with technology: Exploring celestial motion on an interactive whiteboard. *Research in Science Education, 51*(2), 235–275. https://doi.org/10.1007/s11165-018-9794-8.

Guo, H., Li, H., Chan, G., & Skitmore, M. (2012). Using game technologies to improve the safety of construction plant operations. *Accident Analysis and Prevention*. https://doi.org/10.1016/j.aap.2011.06.002.

Gürer, S. (2021). *Development of a Virtual Reality-Based Serious Game for Occupational Health and Safety Training in Underground Mining*. Middle East Technical University,

Hariharasudan, A., & Kot, S. (2018). A scoping review on Digital English and Education 4.0 for Industry 4.0. *Social Sciences, 7*(11). https://doi.org/10.3390/socsci7110227.

Hisham, E.-S., Ghada Ahmed, R., & Amany Ahmed, R. (2015). Using mixed reality as a simulation tool in urban planning project for sustainable development. *Journal of Civil Engineering and Architecture, 9*(7). https://doi.org/10.17265/1934-7359/2015.07.009.

Hou, Y., Jiang, X., & Quan, W. (2019). Construction of virtual simulation experiment teaching system for mining and safety. *Education Modernization, 6*(46), 40–43.

Hou, C., Zhu, W., Liu, H., & Zhang, P. (2021). Teaching reform of mining engineering based on virtual reality technology. *Education and Teaching Forum, 2021*(21), 73–76.

Hsu, T.-C. (2017). Learning English with augmented reality: Do learning styles matter? *Computers & Education, 106*, 137–149. https://doi.org/10.1016/j.compedu.2016.12.007.

Huang, H.-M., Rauch, U., & Liaw, S.-S. (2010). Investigating learners' attitudes toward virtual reality learning environments: Based on a constructivist approach. *Computers & Education, 55*(3), 1171–1182.

Huang, T.-L., & Liao, S. (2015). A model of acceptance of augmented-reality interactive technology: The moderating role of cognitive innovativeness. *Electronic Commerce Research, 15*(2), 269–295. https://doi.org/10.1007/s10660-014-9163-2.

Huang, T.-C., Chen, C.-C., & Chou, Y.-W. (2016). Animating eco-education: To see, feel, and discover in an augmented reality-based experiential learning environment. *Computers & Education, 96*, 72–82. https://doi.org/10.1016/j.compedu.2016.02.008.

Jamhawi, M.M. & Hajahjah, Z.A. (2016). It-innovation and technologies transfer to heritage sites: the case of Madaba, Jordan. *Mediterranean Archaeology & Archaeometry, 16*(2), 41–46.

Janiszewski, M., Uotinen, L., Merkel, J., Leveinen, J., & Rinne, M. (2020). *Virtual Reality Learning Environments for Rock Engineering, Geology and Mining Education.* Paper presented at the 54th US Rock Mechanics/Geomechanics Symposium.

Janiszewski, M., Uotinen, L., Szydlowska, M., Munukka, H., Dong, J., & Rinne, M. (2021). *Visualization of 3D Rock Mass Properties in Underground Tunnels Using Extended Reality.* Paper presented at the IOP Conference Series: Earth and Environmental Science.

Kalkofen, D., Mori, S., Ladinig, T., Daling, L., Abdelrazeq, A., Ebner, M., … Shepel, T. (2020). *Tools for Teaching Mining Students in Virtual Reality Based on 360° Video Experiences.* Paper presented at the 2020 IEEE Conference on Virtual Reality and 3D User Interfaces Abstracts and Workshops (VRW).

Kandi, V., Brittle, P., Castronovo, F., & Gaedicke, C. (2020). Application of a virtual reality educational game to improve design review skills. In *Construction Research Congress 2020*. American Society of Civil Engineers: 545–554.

Ke, F., Lee, S., & Xu, X. (2016). Teaching training in a mixed-reality integrated learning environment. *Computers in Human Behavior, 62*, 212–220. https://doi.org/10.1016/j.chb.2016.03.094.

Kent, L., Snider, C. and Hicks, B. (2019). Engaging citizens with urban planning using city blocks, a mixed reality design and visualisation platform. In *Augmented Reality, Virtual Reality, and Computer Graphics: 6th International Conference*, AVR 2019, Santa Maria al Bagno, Italy, June 24–27, 2019, Proceedings, Part II. Springer International Publishing: 651–662.

Keping Zhou, M. G. (2010). Virtual reality simulation system for underground mining project.

Khan, A., Sepasgozar, S., Liu, T., & Yu, R. (2021). Integration of BIM and immersive technologies for AEC: A scientometric-SWOT analysis and critical content review. *Buildings, 11*(3), 126. https://doi.org/10.3390/buildings11030126.

Kim, Y. M., Rhiu, I., & Yun, M. H. (2020). A systematic review of a virtual reality system from the perspective of user experience. *International Journal of Human–Computer Interaction, 36*(10), 893–910.

Kizil, M. (2003). Virtual reality applications in the Australian minerals industry. *Application of Computers and Operations Research in the Minerals Industries, South African*, 569–574.

Kizil, M. S., Kerridge, A. P., & Hancock, M. G. (2004). Use of virtual reality in mining education and training.

Kourouthanassis, P., Boletsis, C., Bardaki, C., & Chasanidou, D. (2015). Tourists responses to mobile augmented reality travel guides: The role of emotions on adoption behavior. *Pervasive and Mobile Computing, 18*, 71–87. https://doi.org/10.1016/j.pmcj.2014.08.009.

Ku, K., & Mahabaleshwarkar, P. S. (2011). Building interactive modeling for construction education in virtual worlds. *Electronic Journal of Information Technology in Construction*. Retrieved from https://www.scopus.com/inward/record.uri?eid=2-s2.0-79954551580&partnerID=40&md5=df083a55b0a486447cad122a7940be55.

Le, Q. T., Pedro, A., & Park, C. S. (2015). A social virtual reality based construction safety education system for experiential learning. *Journal of Intelligent and Robotic Systems: Theory and Applications*. https://doi.org/10.1007/s10846-014-0112-z.

Lei, B., Wu, B., & Zhou, Y. (2014). Coal mine emergency rescue drill system based on virtual reality technology. *Journal of Chemical and Pharmaceutical Research*. Retrieved from https://www.scopus.com/inward/record.uri?eid=2-s2.0-84907240882&partnerID=40&md5=3abe3e4b449b430513017c9557a5b1d3.

LeMoine, J. (2017). Immersion creating engagement in urban planning and redesign. *IEEE Xplore*.

Lev Rassudov, A. K. (2020). COVID-19 pandemic challenges for engineering education. In *International Conference on Electrical Power Drive Systems*.

Li, H., Chan, G., & Skitmore, M. (2012). Multiuser virtual safety training system for tower crane dismantlement. *Journal of Computing in Civil Engineering*. https://doi.org/10.1061/(ASCE)CP.1943-5487.0000170.

Li, X., Yi, W., Chi, H. L., Wang, X., & Chan, A. P. C. (2018). A critical review of virtual and augmented reality (VR/AR) applications in construction safety. *Automation in Construction*. https://doi.org/10.1016/j.autcon.2017.11.003.

Li, M., Sun, Z., Jiang, Z., Tan, Z., Chen, J., & Chen, C.-H. (2020). A virtual reality platform for safety training in coal mines with AI and cloud computing. *Discrete Dynamics in Nature and Society, 2020*, 1–7. https://doi.org/10.1155/2020/6243085.

Liang, R., Xu, S., Shen, Q., & An, L. (2018). Research and development of dynamic simulation system for mining method based on Unity3D. *Metal Mine, 2018*(02), 141–145.

Liang, R., Xu, S., Hou, P., & Zhu, C. (2019a). Research on 3D modeling technology of mining method for underground mining of metallic deposits. *China Mining Magazine, 28*(03), 73–77.

Liang, Z., Zhou, K., & Gao, K. (2019b). Development of virtual reality serious game for underground rock-related hazards safety training. *IEEE Access*. https://doi.org/10.1109/ACCESS.2019.2934990.

Liljaniemi, A., & Paavilainen, H. (2020). Using digital twin technology in engineering education – Course concept to explore benefits and barriers. *Open Engineering, 10*(1), 377–385. https://doi.org/10.1515/eng-2020-0040.

Lopes, C. V., & Lindström, C. (2012). Virtual cities in Urban planning: The Uppsala case study. *Journal of Theoretical and Applied Electronic Commerce Research*. https://doi.org/10.4067/S0718-18762012000300009.

Madni, A. M. (2019). Exploiting digital twin technology to teach engineering fundamentals and afford real-world learning opportunities. In *2019 ASEE Annual Conference & Exposition*. Association for Computing Machinery.

Mallett, L., & Unger, R. (2010). Virtual reality in mine training. https://www.cdc.gov/niosh/mining%5C/UserFiles/works/pdfs/vrimt.pdf.

Markets, M. a. (2020, September). Digital twin market. Retrieved from https://www.marketsandmarkets.com/Market-Reports/digital-twin-market-225269522.html.

McHaney, R., Reiter Copeland, L., & Reychav, I. (2017). Immersive simulation in constructivist-based classroom E-learning. *International Journal on E-Learning (IJEL), 17*(1), 39–64. Association for the Advancement of Computing in Education (AACE).

Minnery, J., & Searle, G. (2014). Toying with the city? Using the computer game SimCity™4 in planning education. *Planning Practice & Research, 29*(1), 41–55. https://doi.org/10.1080/02697459.2013.829335.

Mitra, R., & Saydam, S. (2014). Can artificial intelligence and fuzzy logic be integrated into virtual reality applications in mining? *Journal of the Southern African Institute of Mining and Metallurgy*. Retrieved from https://www.scopus.com/inward/record.uri?eid=2-s2.0-84961380611&partnerID=40&md5=19605c13b0a9d2da3854d8557251fde1.

Nguyen, M.-T., Nguyen, H.-K., Vo-Lam, K.-D., Nguyen, X.-G., & Tran, M.-T. (2016). Applying virtual reality in city planning. In *Virtual, Augmented and Mixed Reality*. Stephanie Lackey, Randall Shumaker (pp. 724–735). Springer.

Nickel, C., Knight, C., Langille, A., & Godwin, A. (2019). How much practice is required to reduce performance variability in a virtual reality mining simulator? *Safety*. https://doi.org/10.3390/safety5020018.

Nieto-Lugilde, D., Torrecilla-Salinas, C.J., De Troyer, O. and Gutiérrez, J., 2019. Designing a serious game as a tool for landscape and urban planning immersive learning. In *Fifth Immersive Learning ResearchNetwork Conference*. Verlag der Technischen Universität Graz: 140–147.

Nikolaev, S., Gusev, M., Padalitsa, D., Mozhenkov, E., Mishin, S., & Uzhinsky, I. (2018). Implementation of "digital twin" concept for modern project-based engineering education. In *Product Lifecycle Management to Support Industry 4.0: 15th IFIP WG 5.1 International Conference, PLM 2018, Turin, Italy, July 2-4, 2018, Proceedings 15*. Springer International Publishing: 193–203.

Orr, T. J., Mallet, L. G., & Margolis, K. A. (2009). Enhanced fire escape training for mine workers using virtual reality simulation. *Mining Engineering, 61*(11), 41.

Park, C. S., & Kim, H. J. (2013). A framework for construction safety management and visualization system. *Automation in Construction*. https://doi.org/10.1016/j.autcon.2012.09.012.

Pedram, S., Skarbez, R., Palmisano, S., Farrelly, M., & Perez, P. (2021). Lessons learned from immersive and desktop VR training of mines rescuers. *Frontiers in Virtual Reality, 2*. https://doi.org/10.3389/frvir.2021.627333.

Phan, V.T. and Choo, S.Y., 2010. A Combination of Augmented Reality and Google Earth's facilities for urban planning in idea stage. *International Journal of Computer Applications, 4*(3), 26–34.

Piao, H., Duan, H., & Zhu, M. (2020). Simulation of urban landscape around subway station based on machine learning and virtual reality. *Microprocessors and Microsystems*. https://doi.org/10.1016/j.micpro.2020.103495.

Picinali, L., Afonso, A., Denis, M., & Katz, B. F. G. (2014). Exploration of architectural spaces by blind people using auditory virtual reality for the construction of spatial knowledge. *International Journal of Human Computer Studies*. https://doi.org/10.1016/j.ijhcs.2013.12.008.

Poplin, A. (2012). Playful public participation in urban planning: A case study for online serious games. *Computers, Environment and Urban Systems, 36*(3), 195–206. https://doi.org/10.1016/j.compenvurbsys.2011.10.003.

Praveen Maghelal, P. N., Naderi, J. R., & Kweon, B.-S. (2011). Investigating the use of virtual reality for pedestrian environments. *Journal of Architectural and Planning Research. 28*(2), 104–117.

Qiu, X.-y., Chiu, C.-K., Zhao, L.-L., Sun, C.-F., & Chen, S.-j. (2023). Trends in VR/AR technology-supporting language learning from 2008 to 2019: A research perspective. *Interactive Learning Environments, 31*(4), 2090–2113.

Radianti, J., Majchrzak, T. A., Fromm, J., & Wohlgenannt, I. (2020). A systematic review of immersive virtual reality applications for higher education: Design elements, lessons learned, and research agenda. *Computers & Education, 147*, 103778. https://doi.org/10.1016/j.compedu.2019.103778.

Rafsanjani, H. N., & Nabizadeh, A. H. (2021). Towards digital architecture, engineering, and construction (AEC) industry through virtual design and construction (VDC) and digital twin. *Energy and Built Environment*. https://doi.org/10.1016/j.enbenv.2021.10.004.

Reaver, K. (2020). Evaluating the use ofmixed reality(MR) in urban planning. *Proceedings of the eCAADe*.

Redondo, E., Valls, F., Fonseca, D., Navarro, I., Villagrasa, S., Olivares, A., & Peredo, A. (2014). *Educational Qualitative Assessment of Augmented Reality Models and Digital Sketching Applied to Urban Planning*. Paper presented at the Proceedings of the Second International Conference on Technological Ecosystems for Enhancing Multiculturality - TEEM '14.

Redondo, E., Fonseca, D., Sánchez-Sepúlveda, M., Zapata, H., Navarro, I., Gimenez, L., & Pérez, M. A. (2020). EDUGAME4CITY. A gamification for architecture students. Viability study applied to urban design. In *Learning and Collaboration Technologies. Human and Technology Ecosystems*. Ashish GhoshIrwin KingMalay BhattacharyyaShubhra Sankar Ray. Lecture Notes in Computer Science (LNISA, volume 12206). (pp. 296–314.

Sacks, R., Perlman, A., & Barak, R. (2013). Construction safety training using immersive virtual reality. *Construction Management and Economics*. https://doi.org/10.1080/01446193.2013.828844.

Sampaio, A. Z., & Martins, O. P. (2014). The application of virtual reality technology in the construction of bridge: The cantilever and incremental launching methods. *Automation in Construction*. https://doi.org/10.1016/j.autcon.2013.10.015.

Sanchez, B., Ballinas-Gonzalez, R., & Rodriguez-Paz, M. X. (2021, 21–23 April 2021). *Development of a BIM-VR Application for E-Learning Engineering Education*. Paper presented at the 2021 IEEE Global Engineering Education Conference (EDUCON).

Sepasgozar, S. M. E. (2020). Digital twin and web-based virtual gaming technologies for online education: A case of construction management and engineering. *Applied Sciences, 10*(13). https://doi.org/10.3390/app10134678.

Sepasgozar, S. M., Ghobadi, M., Shirowzhan, S., Edwards, D. J., & Delzendeh, E. (2021). Metrics development and modelling the mixed reality and digital twin adoption in the context of Industry 4.0. *Engineering, Construction and Architectural Management 28*(5), 1355–1376.

Shailey Minocha, A. J. R. (2010). *Interaction Design and Usability of Learning Spaces in 3D Multi-User Virtual Worlds*. International Federation for Information Processing.

Shen, Z. (2012). *Geospatial Techniques in Urban Planning*. Springer Science & Business Media. Springer.

Shirowzhan, S., Sepasgozar, S. M., & Trinder, J. (2021). Developing metrics for quantifying buildings' 3D compactness and visualizing point cloud data on a web-based app and dashboard. *Journal of Construction Engineering and Management, 147*(3), 04020178.

Shiva Pedram, P. P., Palmisano, S., & Farrelly, M. (2016). A systematic approach to evaluate the role of virtual reality as a safety training tool in the context of the mining industry.

Song, L., Lu, Z., & Petsangsri, S. (2020). *Reconstruction of Smart Learning Space Based on Digital Twin (DT) Technology*. Paper presented at the 2020 7th International Conference on Dependable Systems and Their Applications (DSA).

Squelch, A. P. (2001). Virtual reality for mine safety training in South Africa. *Journal of the Southern African Institute of Mining and Metallurgy*, *101*(4), 209–216.

Stothard, P., & Laurence, D. (2014). Application of a large-screen immersive visualisation system to demonstrate sustainable mining practices principles. *Mining Technology, 123*(4), 199–206. https://doi.org/10.1179/1743286314y.0000000068.

Stothard, P., Squelch, A., Stone, R., & Van Wyk, E. (2019). Towards sustainable mixed reality simulation for the mining industry. *Mining Technology, 128*(4), 246–254. https://doi.org/10.1080/25726668.2019.1645519.

Tan, B., Zhang, Z., & Qin, X. (2015). Coal and gas outburst accident virtual escape system for miners based on virtools. *Open Automation and Control Systems Journal*. https://doi.org/10.2174/1874444301507010379.

Tang, C., Wang, C., Qu, L., & Ling, L. (2014). Modeling and simulation of virtual environment system of complex coalmine using multi-agent technology. *Environmental Engineering and Management Journal*. https://doi.org/10.30638/eemj.2014.116.

Teizer, J., Cheng, T., & Fang, Y. (2013). Location tracking and data visualization technology to advance construction ironworkers' education and training in safety and productivity. *Automation in Construction*. https://doi.org/10.1016/j.autcon.2013.03.004.

Thurner, S., Daling, L., Ebner, M., Ebner, M., & Schön, S. (2021). Evaluation design for learning with mixed reality in mining education based on a literature review. In *International Conference on Human-Computer Interaction*. Springer International Publishing: 313–325.

Tunçer, B. (2020). Augmenting reality: (Big-)data-informed urban design and planning. *Architectural Design*. https://doi.org/10.1002/ad.2568.

Tunur, T., Hauze, S. W., Frazee, J. P., & Stuhr, P. T. (2021). XR-immersive labs improve student motivation to learn kinesiology. *Frontiers in Virtual Reality, 2*, 15. Retrieved from https://www.frontiersin.org/article/10.3389/frvir.2021.625379.

van Wyk, E., & de Villiers, M. R. (2014). *Applying Design-Based Research for Developing Virtual Reality Training in the South African Mining Industry*. Paper presented at the Proceedings of the Southern African Institute for Computer Scientist and Information Technologists Annual Conference 2014 on SAICSIT 2014 Empowered by Technology - SAICSIT '14.

Van Wyk, E. and De Villiers, R. (2009), February. Virtual reality training applications for the mining industry. In *Proceedings of the 6th international conference on computer graphics, virtual reality, visualisation and interaction in Africa*. Association for Computing Machinery, Inc.: 53–63.

Van Wyk, E. A., & De Villiers, M. R. (2019). An evaluation framework for virtual reality safety training systems in the South African mining industry. *Journal of the Southern African Institute of Mining and Metallurgy*. https://doi.org/10.17159/24119717/53/2019.

Varadarajan, K.M. and Vincze, M., 2011, September. Augmented virtuality based immersive telepresence for control of mining robots. In *2011 5th International Symposium on Computational Intelligence and Intelligent Informatics (ISCIII)*. IEEE: 133–138. https://ieeexplore.ieee.org/xpl/conhome/6062353/proceeding.

Villena Taranilla, R., Cózar-Gutiérrez, R., González-Calero, J. A., & López Cirugeda, I. (2019). Strolling through a city of the Roman Empire: An analysis of the potential of virtual reality to teach history in Primary Education. *Interactive Learning Environments, 30*(4), 608–618.

von der Pütten, A. M., Klatt, J., Ten Broeke, S., McCall, R., Krämer, N. C., Wetzel, R., … Klatt, J. (2012). Subjective and behavioral presence measurement and interactivity in the collaborative augmented reality game TimeWarp☆. *Interacting with Computers, 24*(4), 317–325. https://doi.org/10.1016/j.intcom.2012.03.004.

Wang, M., Wang, C. C., Sepasgozar, S., & Zlatanova, S. (2020). A systematic review of digital technology adoption in off-site construction: Current status and future direction towards industry 4.0. *Buildings, 10*(11), 1–29. https://doi.org/10.3390/buildings10110204.

Webber-Youngman, R. C. W., & Van Wyk, E. A. (2013). Incident reconstruction simulations-potential impact on the prevention of future mine incidents. *Journal of the Southern African Institute of Mining and Metallurgy*. Retrieved from https://www.scopus.com/inward/record.uri?eid=2-s2.0-84882438742&partnerID=40&md5=a234cdf0bab2aa419f618c63b431cdce.

West, M., Yildirim, O., Harte, A.E., Ramram, A., Fleury, N.W. and Carabias-Hütter, V., 2019. Enhancing citizen participation through serious games in virtual reality. In *24th International Conference on Urban Planning, Regional Development and Information Society (REAL CORP 2019), Karlsruche, Germany, 2-4 April 2019*. Competence Center of Urban and Regional Planning: 881–888.

Wojciechowski, R., & Cellary, W. (2013). Evaluation of learners' attitude toward learning in ARIES augmented reality environments. *Computers & Education, 68*, 570–585. https://doi.org/10.1016/j.compedu.2013.02.014.

Xiaoqiang, Z., An, W., & Jianzhong, L. (2011). Design and application of virtual reality system in fully mechanized mining face. *Procedia Engineering, 26*, 2165–2172. https://doi.org/10.1016/j.proeng.2011.11.2421.

Xie, J., Yang, Z., Wang, X., & Wang, Y. (2018a). A remote VR operation system for a fully mechanised coal-mining face using real-time data and collaborative network technology. *Mining Technology: Transactions of the Institute of Mining and Metallurgy*. https://doi.org/10.1080/25726668.2018.1464817.

Yang, T., Zhang, C., Yu, Q., Zhu, W., & Zhang, P. (2019). Construction of the experimental teaching center for virtual simulation of rock mechanics and safe mining in metal mines and studies on teaching practices. *Education Teaching Forum, 2021*(17), 140–145. https://dl.acm.org/doi/abs/10.1145/3473141.3473220.

Yiakoumettis, C., Doulamis, N., Miaoulis, G., & Ghazanfarpour, D. (2014). Active learning of user's preferences estimation towards a personalized 3D navigation of geo-referenced scenes. *GeoInformatica*. https://doi.org/10.1007/s10707-013-0176-0.

Yilmaz, R. M. (2016). Educational magic toys developed with augmented reality technology for early childhood education. *Computers in Human Behavior, 54*, 240–248. https://doi.org/10.1016/j.chb.2015.07.040.

Yitmen, I., Alizadehsalehi, S., Akıner, İ., & Akiner, M. (2021). An adapted model of cognitive digital twins for building lifecycle management. *Applied Sciences, 11*. https://doi.org/10.3390/app11094276.

Zhai, X., Wang, M., & Ghani, U. (2019). The SOR (stimulus-organism-response) paradigm in online learning: An empirical study of students' knowledge hiding perceptions. *Interactive Learning Environments, 28*, 1–16. https://doi.org/10.1080/10494820.2019.1696841.

Zhang, H. (2017). Head-mounted display-based intuitive virtual reality training system for the mining industry. *International Journal of Mining Science and Technology*. https://doi.org/10.1016/j.ijmst.2017.05.005.

Zhang, H., He, X., & Mitri, H. (2019). Fuzzy comprehensive evaluation of virtual reality mine safety training system. *Safety Science, 120*, 341–351. https://doi.org/10.1016/j.ssci.2019.07.009.

Zhang, C., Zhao, Y., Teng, T., Chen, J., & Zhang, L. (2020). Application of mine VR teaching system in the course of Mining Introduction. *Journal of Higher Education*, (04), 99–101.

Zhou, K. and Guo, M., 2011. Virtual reality simulation system for underground mining project. In *Virtual Reality*. IntechOpen.

Zion Market Research. (2019, February 21). Globe news wire. Retrieved from https://www.globenewswire.com/news-release/2019/02/21/1739121/0/en/Global-Augmented-and-Virtual-Reality-Market-Will-Reach-USD-814-7-Billion-By-2025-Zion-Market-Research.html.

6 Digital twin for urban decision support systems

Scientometric and thematic analysis

Peyman Najafi, Ali Soltani, Ayaz Ahmad Khan, Mahsa Chizfahm, Samad M. E. Sepasgozar and Ning Gu

6.1 Introduction

A city digital twin is a computer-generated, dynamic, and interactive virtual model representing the physical urban environment based on real-time data from multiple sources (Sepasgozar, 2020). The bidirectional data flow between the physical and virtual parts of the digital twin allows urban managers, urban planners, and other stakeholders to monitor changes and make informed decisions related to urban services in real-time (Masoumi et al., 2023). Although the terms, urban and city are often used interchangeably, they have subtle differences: urban area covers not only the city but also towns, suburbs, and other forms of developed land. In other words, all cities are urban areas, but not all urban areas are cities (Sepasgozar, 2021). An urban digital twin consists of three key elements: the physical entity, representing the physical urban area in the real world; the digital twin, serving as a virtual or digital replica of the physical entity; and the digital thread, which acts as the communication channel between these two entities. Unlike similar models, digital twins can establish a connection between data, allowing for a two-way integrated data flow between the physical and digital models. This innovative concept sets the stage for exploring the capabilities and applications of digital twins in various fields, such as manufacturing, design, construction, urban planning, and decision-making.

While the initial definitions of digital mirrors and digital twins have existed in the literature for at least a decade, they have recently gained attention as a solution for managing and planning urban built environments (Wu et al., 2023). The expansion of the Internet of Things (IoTs) infrastructure and data hubs, along with advancements in 3D modelling, artificial intelligence (AI), and visualisation techniques, make digital twins particularly attractive for application in urban contexts (Ketzler et al., 2020). Every object or entity in an urban digital twin may have a traceable historical record, a current status that can be examined, and a predictable future state. This feature makes the digital twin an essential driving force for urban intelligence. Decision-makers can accomplish better organised and efficient urban governance with the availability of processed data, and citizens can monitor and engage in urban governance procedures (Deng et al., 2021).

DOI: 10.1201/9781003507000-6

Digital urban planning has shown potential benefits that can be obtained in sustainability, efficiency, optimisation, collaboration, and innovation. However, adoption remains low, and many questions remain about integrating it into a cohesive and adaptable framework compatible with urban environments. Furthermore, it can be a complex and expensive practice to build a comprehensive digital twin model of an entire city, including all the objects and items within it, exchanging large amounts of data, and storing the data on the cloud in convenient ways.

Despite these challenges, there is an untapped potential associated with the knowledge gained from digital twin projects, which could beneficially inform future urban planning efforts. Therefore, this article aims to identify the potential of digital twins in the urban context, highlight the challenges that may impede their use, and propose a research agenda for their future development. The research agenda presented in this chapter is critical in maximising the utilisation of digital twins in enhancing awareness of future urban activities, processes, and trends and improving urban management, all of which can have significant benefits at the city, national, and international levels.

The chapter is structured as follows: in the second section, we present the methods used to collect and categorise practical studies in this field. In the third section, we provide a thematic classification and detailed explanation of the potential and challenges of digital twins, as well as the level of development achieved thus far. Finally, in the concluding section, we present a discussion and proposed research agenda.

6.2 Research objective and method

This chapter aims to develop a comprehensive research agenda that identifies the potential and challenges of implementing digital twins in urban contexts. The agenda will also explore the level of maturity of existing digital twin projects across various fields, as well as the critical requirements for future research on this emerging concept in urban planning.

To achieve this objective, a systematic review was conducted using a range of databases and search engines, including Scopus, Web of Science, Google Scholar, and a snowballing hand search. The review focused on English language scholarly journals published from January 2021 onwards, using the search term "digital twin" in combination with the words "city," "urban," "town," "planning," and "built environment." Non-peer-reviewed publications were excluded, and review papers were used as "grey literature" to supplement hand-selected relevant resources. The final articles' screening and selection process was conducted as per the PRISMA guidelines (Moher et al., 2009) (Figure 6.1).

To identify the current research directions in digital twins in urban contexts, the study conducted a thematic and scientometric analysis of selected articles using the qualitative data analysis method. This combinatory method involves identifying patterns in a dataset and constructing a coupling network based on qualitative content relationships.

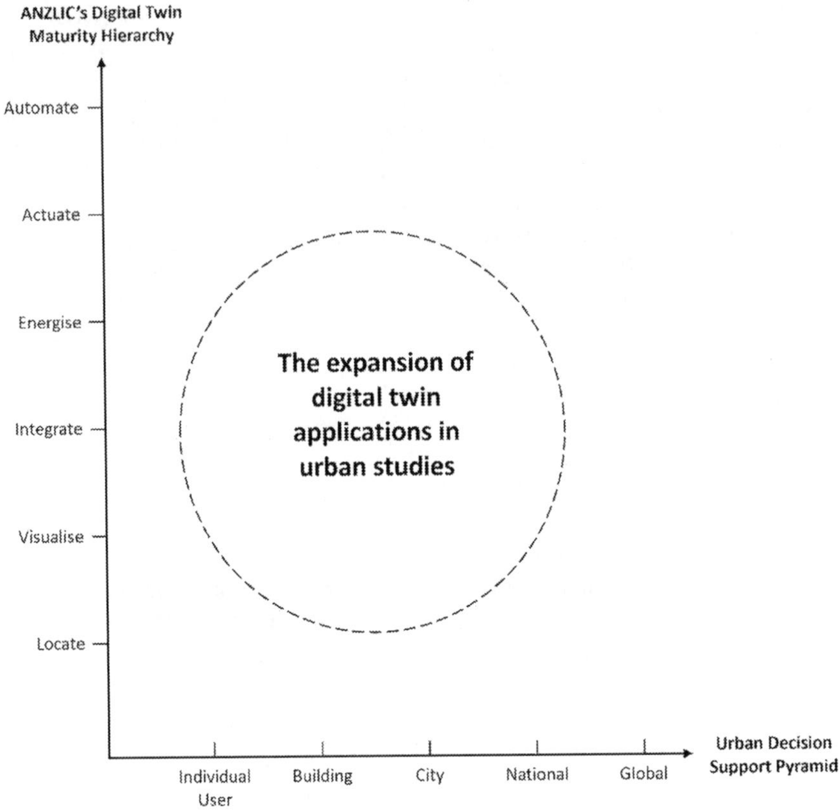

Figure 6.1 The ANZLIC-UDS exploration framework. The *Y*-axis signifies the degree of advancement and contribution of digital twin applications within urban studies. The *X*-axis determines the urban decision-making scale at which digital twins support decisions. The dashed circle schematically denotes that the scope of the studies is open-ended and may extend beyond the current framework.

Source: https://www.anzlic.gov.au/ and (ANZLIC, 2019).

Thematic analysis is an interpretive method that involves constructing threads or patterns from a dataset to address a research question (Vaismoradi et al., 2016). In digital twin research, the term "thread" is used instead of "theme" to encompass the three key elements of the digital twin model: the physical entity, the digital counterpart, and the digital thread. The study utilised tools like VOS Viewer to conduct keyword analysis of the selected articles and assess the maturity level of digital threads relevant to our research interest.

The scientometric analysis involves constructing coupling networks based on scholarly journal publications' titles, abstracts, and keyword relations, as shown in Appendix A and Table 6.1. These networks enable extracting and visualising co-occurrence networks of essential threads from the digital twin and urban planning literature. Visualising the relationships between different threads can gain deeper insights into the complex interconnections and interactions within the literature and highlight areas where further research is needed.

After identifying the research directions, we employed the Australian and New Zealand Land Information Council (ANZLIC) 6-stage maturity model (ANZLIC, 2019) to assess the extent and manner in which digital twins contribute to our identified research directions. The ANZLIC maturity model serves as a systematic approach to determine the readiness and potential benefits of digital twins within the field of built environment studies (ANZLIC, 2019). Essentially, this model is rooted in the UK's Centre for Digital Built Britain's Gemini Principles, which articulates a vision of a federated ecosystem of digital twins and their benefits for the societal economy.

The ANZLIC model consists of six stages for the maturation of digital twins: locate, visualise, integrate, energise, actuate, and automate. The locate stage establishes the foundational element by offering authoritative spatial data that organises and structures subsequent stages. The visualise stage allows for the inspection, planning, engagement, and modelling of 2D maps or 3D models. By integrating the model with static data, metadata, and building information modelling (BIM), the integrate stage facilitates data-driven decision-making for efficient asset management (Dani et al., 2023). The energise stage involves real-time data integration enabled by IoT and sensor networks. The actuate stage involves two-way data integration and interaction, and the automate stage utilises artificial intelligence, real-time data, and advanced analytics to automate operations and maintenance.

After elucidating the extent and manner in which digital twins contribute to our identified research direction, we employed the urban decision support (UDS) pyramid to illuminate how our findings align with the goals and policies pertaining to urban planning and development decisions. The UDS pyramid, primarily developed by the Global Development of Research Centre of Japan (Srinivas, 2015), now serves as a comprehensive guide for prioritising and organising decisions within the built environment, urban planning, and development. This pyramid encompasses five levels of decision-making: individual user, building, city, national, and global. At the individual user level, decisions prioritise the needs and preferences of individual users, such as housing, transportation, and public services. The building level encompasses decisions related to the design, construction, and operation of buildings within the urban environment, including energy efficiency, sustainability, and safety. At the city level, decisions focus on the overall development and management of the urban environment, including land use, transportation, infrastructure, and public services. The national level encompasses decisions about national policies and regulations concerning urban development, such as economic development, social equity, and environmental sustainability. Lastly, at the global level, decisions prioritise the broader international context and include decisions concerning global agreements and policies related to urban development and environmental sustainability.

In our analytical approach, we seamlessly integrate the ANZLIC maturity model and the UDS pyramid. The ANZLIC maturity model is a valuable framework for assessing and uncovering the potential applications and implications of digital twins in urban contexts. Subsequently, the employment of the UDS pyramid helps structure and prioritise our assessment of urban decision-making support systems. Using this integrative approach, we aim to contribute to the broader academic and practical discourse on the integration of digital twins into urban planning and development decision-making. Figure 6.1 illustrates the exploration framework of the ANZLIC-UDS model for our thematic and scientometric analysis.

6.3 Results

6.3.1 Summary of results

Three search engines initially obtained 140 journal articles (published between January 2019 and May 2022). However, after title and abstract screening, 75 articles were excluded. The reasons for exclusion included: (1) non-peer-reviewed papers and review articles, (2) lack of focus on urban context, (3) discussing digital twins only as a future method, (4) investigating digital twins on scales different from the UDS levels, and (5) the studies propose frameworks without testing them in a case study.

Further analysis of the remaining 65 articles revealed that 21 did not focus on executing the urban digital twin, resulting in their exclusion. As a result, 44 articles were included in the final review process. Appendix A provides a detailed table for the included articles, and Figure 6.2 provides an overview of the selection process.

6.3.2 Scope of the reviewed studies

The thematic analysis reveals a marked uptick in the number of peer-reviewed articles focused on digital twin and their applicability to urban planning and related decision-support systems. Before 2019, the scholarly literature on this domain was limited to a few publications. Between 2021 and May 2022, 44 peer-reviewed articles experimentally examined the ability of digital twins to support decisions and policies related to urban issues and processes (excluding review articles and non-case studies). Notably, 59% of these articles were published in 2021, with 41% released in the first four months of 2022, attesting to the field's burgeoning growth.

The 44 selected articles were published in 38 diverse journals, covering a broad spectrum of themes ranging from sustainability to water management, energy efficiency, construction, transportation, economics, and more. This diversity speaks to the versatility of the digital twin's potential for application in various urban domains. MDPI, Elsevier, and IEEE emerged as the top publishers in this field, publishing 14, 11, and nine articles, respectively, reflecting their growing role in this research domain. Table 6.1 provides an overview of the distribution of peer-reviewed papers on digital twins in UDS systems across various publishers and journals.

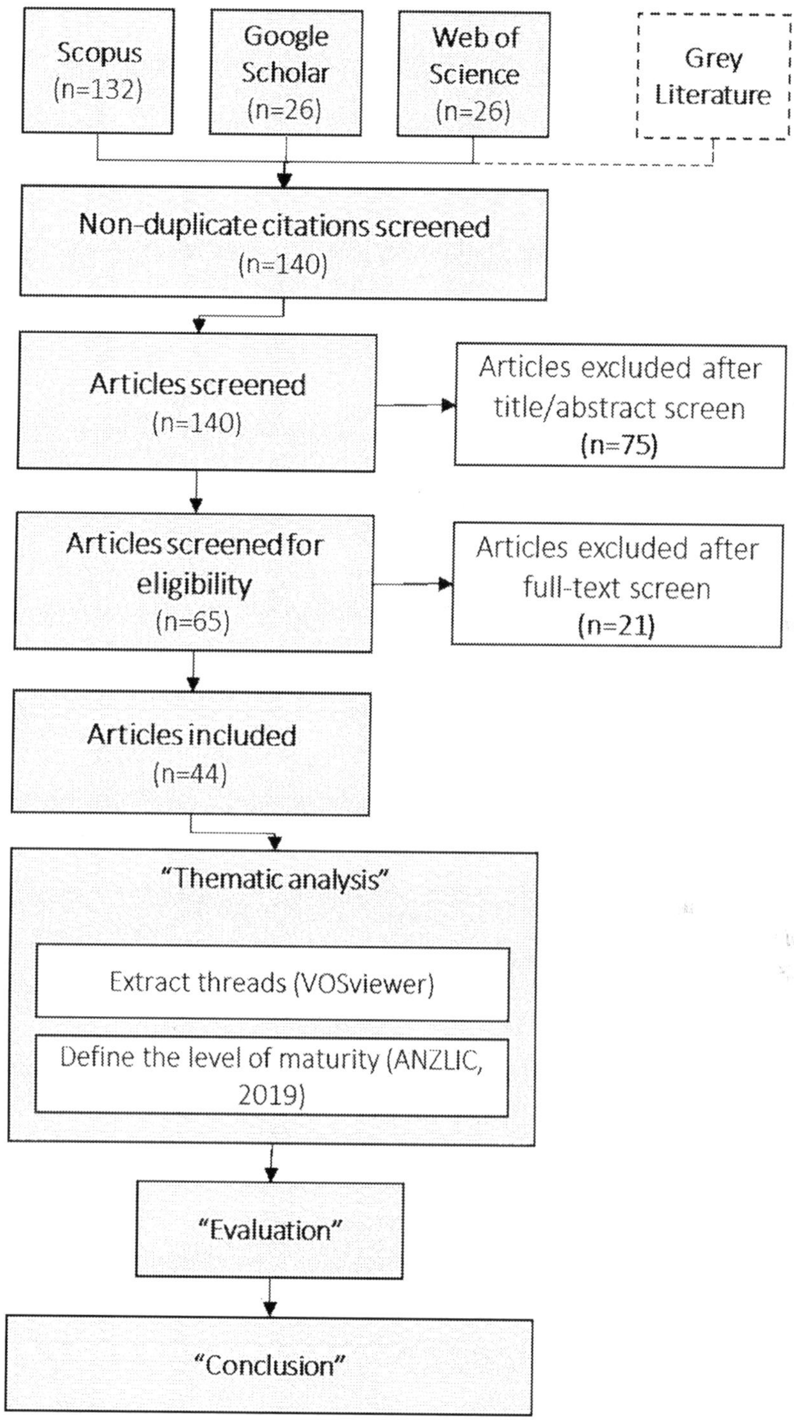

Figure 6.2 The article selection flowchart based on the PRISMA protocol and the papers identified for the present review.

Table 6.1 Distribution of the identified peer-reviewed papers in various journals

Publisher title	*Journal title*	*Journal frequency*	*Publisher frequency (%)*
MDPI	Sustainability	3	32
	Water	2	
	Sensors	2	
	Remote Sensing	2	
	Energies	1	
	Applied Sciences	1	
	Environmental Sciences Proceedings	1	
	Computational Social Science	1	
	International Journal of Geo-Information	1	
Elsevier	Sustainable Cities and Society	2	27
	Cities	1	
	Automation in Construction	1	
	Energy & Buildings	1	
	Environmental Modelling and Software	1	
	Future Generation Computer Systems	1	
	International Journal of Production Economics	1	
	Transportation Engineering	1	
	Transportation Research Part C	1	
	Expert Systems with Applications	1	
IEEE	Access	1	21
	Instrumentation & Measurement Magazine	1	
	Internet Computing	1	
	Intelligent Environments	1	
	Network	1	
	Transactions on Industrial Informatics	1	
	Transactions on Emerging Topics in Computing	1	
	Transactions on Cloud Computing	1	
	Tsinghua Science and Technology	1	
Springer	European Transport Research Review	1	2
IxD&A	Interaction Design and Architecture(s)	1	2
Cambridge Core	Data-Centric Engineering	1	2
Frontiers	Frontiers in Plant Science	1	2
Hindawi	Complexity	1	2
SAGE	International Journal of Architectural Computing	1	2
Taylor & Francis	Urban Technology	1	2
WIT Press	Energy Production and Management	1	2
ASCE	Transportation Engineering, Part A: Systems	1	2
Oxford Academic	Journal of Computational Design and Engineering	1	2

Table 6.2 Distribution of peer-reviewed articles concerning countries and the global context

Global	*Country*	*Country frequency*	*Global frequency (%)*
Global North	UK	7	82
	USA	5	
	Spain	3	
	Canada	3	
	Korea	3	
	Germany	2	
	Belgium	2	
	Ireland	2	
	Italy	2	
	Greece	2	
	Australia	2	
	Finland	1	
	Norway	1	
	Denmark	1	
	The Netherlands	1	
	Sweden	1	
	Czech Republic	1	
	Estonia	1	
Global South	China	7	18
	Singapore	1	
	South Africa	1	

6.3.3 *The global contribution of the studies*

The thematic analysis brought to light the disproportionate distribution of the studies across the Global North and Global South. In the final selection of 44 papers, 82% of the research was conducted in countries in the Global North. In contrast, only 18% of the papers originated from Global South countries, including China, Singapore, and South Africa. Notably, the countries that made the most significant contributions to this field of study are the United Kingdom, the USA, and China, among others.

Table 6.2 provides a comprehensive overview of the frequency of peer-reviewed articles based on their respective countries and global contexts. It suggests an uneven development and deployment of digital twins in UDS systems between the Global North and Global South.

6.3.4 *Thematic analysis: discovering the threads*

The study utilised VOS Viewer analytical software to assess the interrelationships between the extracted keywords from the databases. The analysis resulted in the identification of 89 items from the 44 citations, which were then grouped into five distinct threads, including "Urban Information Systems and Management (UI&S)," "Urban Energy (UE)," "Urban Mobility (UM)," "Urban Community Participatory Management (UCPM)," and "Urban Modelling and Visualisation (UM&V)" (Figure 6.3).

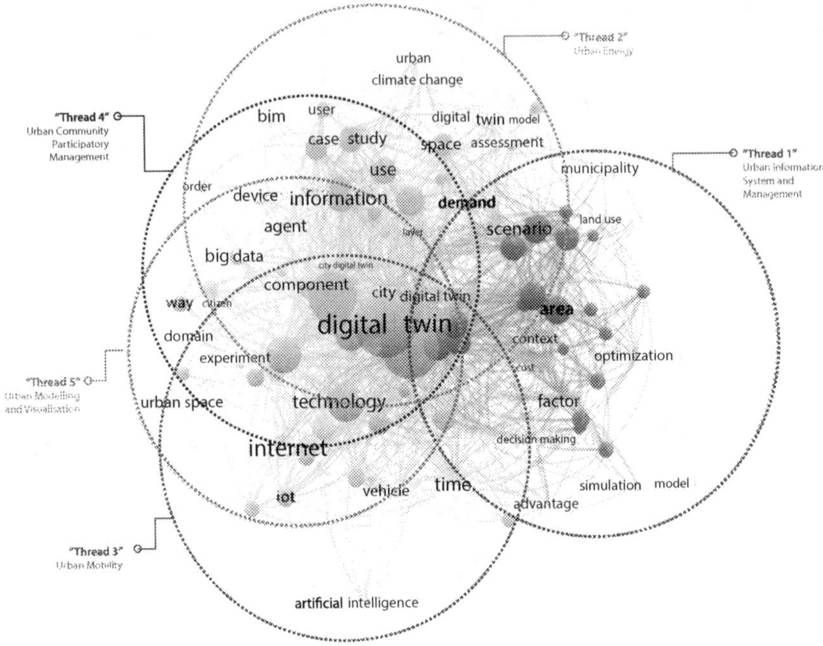

Figure 6.3 Map of the threads extracted from articles on utilising digital twins in urban decision support systems.

1 UI&S encompasses 30 items, with "framework," "decision" (encompassing decision-making and decision-maker), "scenario," and "implementation" constituting the most salient keywords.
2 UE comprises 21 items: "BIM," "climate change," "energy," "information," "big data," and "assessment."
3 UM comprises 16 items, the most critical being "smart cities," "IoT," and "real-time."
4 UCPM encompasses 14 items: "citizen," "experiment," "interaction," "real-world," and "augmented reality."
5 UM&V comprises eight items, with recurrent keywords such as "data," "concept," "knowledge," and "agent."

6.3.4.1 *Thread 1: Urban Information Systems and Management (UI&S)*

Thirteen articles are showcased in this thread, highlighting that digital twins are conducted at three different urban scales ranging from building (urban block), to city (part of), and nationwide (Table 6.3). In practice, most of the digital twin applications are devoted to the city level, representing 77% of the corpus. On the other hand, its application at the national scale accounts for 15%, while the building scale only comprises 8% of the studies within this thread (Table 6.3).

It is worth noting that the articles in this thread display an impressive degree of maturity, with more than half (54%) of the digital twin applications attaining

Table 6.3 Reviewed papers in UI&S, highlighting their study scale, digital twin application level, and focus area

References	*Study scale*	*Digital Twining maturity level*	*Focus area*
Truu et al. (2021)	City	Integrate	Flood-resilience
Gürdür Broo et al. (2022)	Building	Actuate	Smart infrastructure
Bartos and Kerkez (2021)	City	Actuate	Drainage systems
Pedersen et al. (2021)	City	Energise	Urban water systems
Leung et al. (2022)	City	Actuate	Urban distribution system
Bujari et al. (2021)	City	Actuate	Urban facility management
Gutierrez-Franco et al. (2021)	National	Energise	Urban distribution system
Ghandar et al. (2021)	City	Actuate	Urban agriculture
Li et al. (2022)	City	Actuate	Urban governance
Raes et al. (2022)	City	Actuate	Smart city
Badawi et al. (2021)	City	Integrate	City services
Pang et al. (2020)	National	Automate	Collaborative city
Zhao et al. (2022)	City	Integrate	Urban expansion and vegetation coverage

the actuating level. Furthermore, 23%, 15%, and 8% of the applications exhibit integrated, energised, and automated levels, respectively, demonstrating the wide-ranging methods researchers use to confront the challenge of digital twinning across various urban decision-making scales.

6.3.4.1.1 THE BENEFITS OF DIGITAL TWINS IN UI&S

- Flexibility in urban governance: Digital twin offers flexibility by enabling decision-makers to experiment with different scenarios and make informed choices based on real-time/historical data. This application can create a digital model of the urban environment by collecting data from IoT devices and public databases. Moreover, machine learning (ML) and AI techniques can facilitate remote decision-making and provide performance insights on physical assets. Recent research has emphasised the integration of AI and ML with digital twins to enhance decision-making. By leveraging historical data and cognitive performance control, these techniques can automatically make predictions to enable optimal responses without requiring human intervention.
- Time-saving: Adopting digital twins is a game-changing approach to saving time by automating complex calculations and analysis. As a result, it can facilitate a more efficient and accurate evaluation of scenarios within a shorter timeframe. For example, Li et al. (2022) suggested a novel deep learning (DL)-based digital twin model for cities that exhibit a predictive accuracy of 97.8% by reducing data transfer delays. This study exemplifies the promising advantages of digital twins in revolutionising decision or policy-making methods, ultimately leading to enhanced efficiency and accuracy.

- Logistics and last-mile delivery: With the rise of e-commerce and the effects of the COVID-19 pandemic, digital twins and IoT can be integrated into warehouse systems to minimise the human effort to replenish stocks. Significantly, the integration of advanced AI and the IoT, ensures optimal operational capacity and reduces costs, greenhouse gas emissions, and time in the supply chain (Gutierrez-Franco et al., 2021). Similarly, researchers have proposed a digital twin-based framework to streamline operations in hyper-connected city logistics systems, providing the global movement of physical objects. A novel approach known as the physical internet (PI) shifts closed and independent logistics networks to an open and efficient network that enables rapid delivery of tangible goods to a predetermined location (Gürdür Broo et al., 2022; Leung et al., 2022).
- Resilience: The digital twin has emerged as a promising application for bolstering the resilience of cities against natural disasters and other hazards. It identifies areas with varying degrees of risk, empowering urban decision-makers to proactively address potential disasters in the present and future. For example, in water and flood management, the digital twin has proven instrumental in enabling managers to comprehend the dynamics of urban water systems, minimising the risk of floods during the planning process (Bartos & Kerkez, 2021). In extreme weather planning it also facilitates forecasting of how different spatial planning scenarios may impact the risk of flooding (Truu et al., 2021).

  In addition to disaster planning, the interest in sharing experiences of digital urban modelling among cities is also growing, especially in response to the challenges posed by the COVID-19 pandemic (Pang et al., 2020). The sharing of firsthand knowledge within networks plays a crucial role in preventing planning mistakes from being replicated in different cities. This, in turn, facilitates the provision of optimised decisions in the shortest possible time, ultimately enhancing the resilience of cities (Pang et al., 2020). However, it is worth noting that the potential of digital twins in this regard is still in the early stages of implementation.
- Human-centric planning: The concept of human-centric digital twins is an emerging area of interest in urban studies (Ye et al., 2022). This method promotes considering citizens' behaviour and emotional states when creating digital models of cities. To successfully execute this idea, a human-centric digital twin that uses biometric data, pedestrian models, and tracking or sensing tools to monitor human interactions with their surroundings in real-time, should be created (Raes et al., 2022). Without the need for surveys or research, integrating the user-centric component into a digital twin architecture provides the opportunity to quickly assess the effects of small changes on the user experience (Najafi et al., 2023). This approach advocates consideration of citizens' behaviour and state when developing urban digital models. Successful implementation of this concept involves the creation of a human-centric digital twin that captures citizens' real-time interaction with their environment through biometric data, pedestrian models, and other tracking technologies (Raes et al., 2022). Additionally, digital twins offer a platform for stakeholder engagement around

common problems, encourage consensus building, and transparently test solutions before implementation (Najafi et al., 2022).

Consequently, digital twins serve as a civic technology that enhances public participation (Sepasgozar, 2020). Moreover, citizen-centric digital twins can improve service delivery while minimising operational conflicts and costs (Bujari et al., 2021). This is achieved by encouraging informed planning of operations that account for stakeholder involvement and potential impacts. Overall, human-centric digital twins present a promising approach to urban planning that offers opportunities to improve service delivery, reduce operational conflicts and costs, and enhance public participation in decision-making processes.

6.3.4.1.2 LIMITATIONS AND CHALLENGES OF DIGITAL TWINS IN UI&S

- Security and reliability: Digital twins are susceptible to cyber-attacks, compromising sensitive data and impeding the advancement of public participation objectives and real-time data collection (Raes et al., 2022). Therefore, to mitigate these concerns, they must adhere to rigorous privacy and security standards, such as the European Union's General Data Protection Regulation (GDPR). It is suggested that integrating ML into digital twins may help overcome security and privacy challenges. However, federated learning (FL) techniques are still required to share parameters and maintain privacy while mitigating security risks (Pang et al., 2020).
- Scalability: Implementing a comprehensive digital twin at the city scale poses a significant challenge due to the complexity of cities. They are intricate systems that are challenging to simulate in full detail (Najafi et al., 2023). A suggested approach involves a phased implementation, starting with developing the digital twin for a priority urban area, such as a central business district, and gradually expanding over time. This strategy may allow for the creation of a collection of interconnected digital twins that optimise the performance of organisations both within cities and beyond (Wang et al., 2023).
- Organisational integration: Successful implementation of digital twin models requires adopting organisational practices that bridge knowledge gaps, promote mutual understanding, and foster collaborative and multidisciplinary approaches (Gürdür Broo et al., 2022). Digital twins can provide opportunities for organisational integration beyond incremental technological advancements. Nevertheless, a systemic perspective is critical when designing and implementing them for urban initiatives. Systemic information and organisational perspectives are the primary areas to consider when implementing urban digital twins.

6.3.4.2 Thread 2: Urban Energy

The Urban Energy (UE) thread includes six studies (Table 6.4) that address urban energy-sensitive decision-making at two levels of building and in parts of the city. The reviewed studies suggest that the research related to urban energy is still in the early stages of implementing digital twins, with 67% of the research at the

Table 6.4 Reviewed papers in UE, highlighting their study scale, digital twin application level, and focus area

References	*Study scale*	*Digital Twining maturity level*	*Focus area*
Akroyd et al. (2022)	City	Integrate	Description of landuse
Buckley et al. (2021)	City	Integrate	Energy-resilient
Orozco-Messana et al. (2021)	Building	Energise	Building certification
Marchione and Ruperto (2022)	City	Integrate	Energy distribution
HosseiniHaghighi et al. (2022b)	City	Integrate	Housing energy modelling and mapping
HosseiniHaghighi et al. (2022b)	City	Energise	District heating load

integrative level and 33% at the energising level. Even at the building level, as indicated in (Orozco-Messana et al., 2021), much work still needs to be done to implement the actuate and automate stages of digital twins.

6.3.4.2.1 THE BENEFITS OF DIGITAL TWINS IN UE

- Enhancement in energy distribution networks: Digital twins can enhance flexibility in energy distribution networks by leveraging real-time simulation methods. In particular, digital twins can identify optimal approaches for extracting and redistributing energy based on buildings' spatial and temporal supply and demand. For example, Buckley et al. (2021) employed a digital twin model to manage energy demand at the neighbourhood level, allowing the testing of multiple energy scenarios designed for a neighbourhood, which identified the best approach to extract and redistribute energy based on the temporal and spatial supply and building demand. Similarly, Marchione and Ruperto (2022) and HosseiniHaghighi et al. (2022b) proposed a digital twin model that enhanced the flexibility of energy distribution systems by providing accurate insights into energy supply and demand at the community level.
- Increasing the accuracy of energy simulations: Research indicates a substantial disparity between the predicted energy consumption of a building, as calculated by conventional simulation models, and its actual energy consumption in the real world. However, integrating digital twins with real-time data has shown a remarkable enhancement in the accuracy of energy simulations. For instance, in an experimental study, HosseiniHaghighi et al. (2022b) demonstrated that the average intensity of simulated heating energy consumption in a digital twin model deviated by only 2.5% from actual data in a case study located in British Columbia. Similarly, Orozco-Messana et al. (2021) observed in a study conducted in a neighbourhood in Valencia, Spain, that the calculated energy consumption using their proposed digital twin model had less than a 10% error

compared to the actual consumption for a similar period. Given these findings, the overarching assumption is that a precise, hybrid, or multi-data workflow through digital twin models can provide energy simulations that closely align with reality.

- Attaining sustainable development goals and carbon reduction: Studies have highlighted the effectiveness of using digital twin models to assess the impact of sustainability policies on meeting climate change goals. Orozco-Messana et al. (2021) proposed using a combination of building parameters and real-time data to evaluate urban sustainability, while HosseiniHaghighi et al. (2022a) suggested using digital twins of cities for energy mapping and resilience development. Buckley et al. (2021) used an urban modelling interface to evaluate policies for reducing carbon emissions. These studies indicate that digital twinning can accurately analyse the impact of multiple factors on energy, such as building form, climate, and weather conditions, and support flexible and optimised policymaking, helping to achieve sustainable development goals.

6.3.4.2.2 LIMITATIONS AND CHALLENGES OF DIGITAL TWINS IN UE

- Frequency of data and data processing methods: The primary challenge in the practical implementation of digital twin models in UE is the frequency of data points, which often suffers from heterogeneous data coverage. This issue has been highlighted in 80% of the studies surveyed (Akroyd et al., 2022; Buckley et al., 2021; HosseiniHaghighi et al., 2022b; Marchione & Ruperto, 2022). Despite several varied databases, there is little coherence between data sources in different cities and countries, leading to difficulties in establishing a uniform pattern on large scales between cities.
- Limited access to real-time data at the implementation scale: Studies indicate that incorporating real-time or near real-time data into calculations at the implementation scale has not been achieved to any degree, and numerous challenges remain. Only 33% of the reviewed articles successfully incorporated real-time data and reached the energise level. It is important to note that real-time data is not limited to environmental information, such as weather conditions. For example, HosseiniHaghighi et al. (2022b) observed that even minor building permit interventions are immediately recorded (i.e., in real time), but this information is not updated in the digital twin model. It is essential to continuously integrate real-time or near real-time data from building permits and the digital twin model to optimise and enhance the efficiency of municipal decisions.

6.3.4.3 Thread 3: Urban Mobility

Table 6.5 summarises 13 papers in Urban Mobility (UM), including their references, scales, maturity levels, and focus areas. These papers collectively indicate a wide-ranging application of digital twins in UM studies, encompassing travel demand, traffic analysis, safe driving, and smart traffic flow prediction. Notably, many of these applications demonstrate a high level of maturity. This observation

Table 6.5 Reviewed papers in UM, highlighting their study scale, digital twin application level, and focus area

References	*Study scale*	*Digital Twining maturity level*	*Focus area*
Anda et al. (2021)	City	Energise	Travel demand
Broekman et al. (2021)	Road	Automate	Traffic analysis
Charissis et al. (2021)	National	Automate	Safe driving
Lenfers et al. (2021)	City	Actuate	Traffic models' predictive capabilities
Major et al. (2021)	City	Actuate	4D visualisation for transparency and awareness
Marai et al. (2021)	Road	Automate	Roads infrastructure
Saroj et al. (2021)	Road	Energise	Traffic simulation
Hu et al. (2022)	City	Energise	Smart traffic flow and velocity prediction
Jiang et al. (2022)	City	Energise	Road planning
Lee et al. (2022b)	City	Actuate	Large-scale individual mobility (vehicles and pedestrians) simulation
Argota Sánchez-Vaquerizo (2021)	National	Actuate	Large-scale traffic microsimulation
Steinmetz et al. (2022)	City	Automate	Car-as-a-Service (CaaS)
Van de Vyvere and Colpaert (2022)	City	Energise	ANPR data as traffic analysis tools

suggests a significant potential for the practical application of digital twins in advancing UM studies.

6.3.4.3.1 EXPLORING THE BENEFITS OF DIGITAL TWINS IN UM

- Simplification of traffic analysis: The development of IoT, unmanned aerial vehicles (UAVs), AI, and low-power computing has enabled real-time data access, making it easier to analyse and manage urban traffic. The Internet of Vehicles (IoVs) generates substantial real-time traffic data, which can be input for digital twin models. By connecting physical vehicles and their virtual representation through 5G communications, traffic managers can optimise traffic scheduling and reduce traffic by analysing digital twin traffic data. Because microsimulation enables the exploration of alternative scenarios that might not be tested in real life, it is also beneficial for educating citizens, policymakers, urban planners, and researchers.
- Large-scale agent-based traffic microsimulation: Agent-based traffic microsimulations can employ a stochastic adaptive routing technique and incorporate traffic dynamics, leading to the development of comprehensive and realistic 24-hour digital twin models (Argota Sánchez-Vaquerizo, 2021). For example, the digital twin proposed by Broekman et al. (2021) and Argota Sánchez-Vaquerizo

(2021) employs large-scale microsimulation, accurately counting and classifying vehicles while demonstrating strong alignment with real-world conditions. This approach opens up new operational collaboration opportunities and effectively informs urban policy.

- Environmental monitoring and insights: Digital twins can provide monitoring and insights into the environmental impacts of smart transportation systems. The digital twin was able to simulate the present traffic situation and offer dynamic feedback on environmental performance metrics like energy consumption and vehicle emissions, according to a study by Saroj et al. (2021). Steinmetz et al. (2022) propose a Car-as-a-Service model to promote car-sharing as a public service, which can change how people view and use cars and reduce vehicle usage.
- Increased safety: Digital twins can lead to the development of safer driving infrastructure. Charissis et al. (2021) present a new AR-based approach that visually displays information through a heads-up display and allows interaction with the road system. This model's laboratory testing results show a 64% improvement in accident prevention as users focus their attention on the car's windshield. Additionally, Marai et al. (2021) have designed an automated digital twin model with high accuracy and independent analysis capabilities by proposing the deployment of digital twin boxes on roads. This model can provide safe driving for self-driving vehicles in the future.

6.3.4.3.2 LIMITATIONS AND CHALLENGES OF DIGITAL TWINS IN UM

- Simulating the behaviour of vehicles and citizens in the extensive network of mobility systems: Accurately simulating the behaviour of cars and pedestrians in the network is critical for creating accurate digital twin models. While previous studies have successfully simulated 3D topography, climate conditions, and real-time road traffic, simulating moving data such as vehicles and pedestrians remains difficult. Researchers have proposed using cameras in public places to provide accurate simulations; however, creating dynamic simulation models based on real-world parameters remains a complex problem. Consequently, incorporating human behaviour, which is characterised by awareness, creativity, and sociability, into these simulations remains a challenging task (Anda et al., 2021; Lee et al., 2022a; Marai et al., 2021; Steinmetz et al., 2022; Van de Vyvere & Colpaert, 2022).
- Preserving information security and user privacy: This issue is crucial when using mobile phone data and street camera images to enhance travel demand models and transportation management. Researchers have proposed anonymising data from automatic number plate recognition (ANPR) cameras using histograms provided by telecommunication service providers (TSPs). These cameras may help create personalised travel demand frameworks using blockchain-based solutions (Anda et al., 2021; Lee et al., 2022a; Marai et al., 2021; Steinmetz et al., 2022; Van de Vyvere & Colpaert, 2022). Nonetheless, the efficacy of such cameras still necessitates further research.

- Comprehensibility: Big data can be overwhelming for non-experts, hindering their ability to extract meaningful insights. To address this challenge, some studies have suggested using high-quality 3D graphical digital twins (GDTs) to create 4D visualisations from geo-localised time-series data. However, the widespread accessibility and cost-effectiveness of GDTs is still an open question, as noted by (Major et al., 2021).
- Scalability: Current modelling tools may not be able to handle the complexity and scale of real-world applications, particularly those with large scales and multiple interrelated components. This has been highlighted in several studies, including (Saroj et al., 2021; Steinmetz et al., 2022), which have emphasised the need for more advanced and scalable modelling tools to represent complex systems in the real world accurately.
- Real-time data operational capacity: The current operational capacity of advanced digital twin architecture in real-time is limited, leading to significant delays due to data aggregation. Further research is required to explore the cited solutions, such as using dynamic local maps and registering data for each vehicle to achieve real-time simulated network placement of vehicles.
- Infrastructure placement: The placement of digital twin boxes (DTBs) is critical in reducing deployment and maintenance costs while guaranteeing adequate road coverage. The optimal placement of DTBs may help to minimise costs, ensure that the digital twin models accurately reflect the physical environment, and provide valuable insights for decision-making processes. This has been emphasised by (Marai et al., 2021) in their research.

6.3.4.4 Thread 4: Urban Community Participatory Management

Table 6.6 summarises six studies examining the use of digital twins in Urban Community Participatory Management (UCPM). The table reveals that four of the six studies still need to implement digital twins and are still in the early formulation stages. In contrast, two studies have reached a more mature level in actual implementation and integration. The focus areas of these studies encompass citizen feedback, community participation, transparency, socio-technical perspectives, and universalism. Notably, the two studies that have reached a higher level of maturity in digital twinning both focus on community participation, suggesting that this aspect may be crucial for future research.

6.3.4.4.1 EXPLORING THE BENEFITS OF DIGITAL TWINS IN UCPM

- Empowering agent participation: Digital twins can potentially empower residents and stakeholders to participate in urban decision-making (Najafi et al., 2022). Empowering is achieved through reactive and predictive participation approaches. Reactive participation involves real-time or near real-time interventions to improve urban processes in the short term, while predictive participation focuses on long-term decision-making and urban management through scenario planning (Charitonidou, 2022). For instance, White et al. (2021) demonstrated

Table 6.6 Reviewed papers in UCPM, highlighting their study scale, digital twin application level, and focus area

References	*Study scale*	*Digital Twining maturity level*	*Focus area*
White et al. (2021)	City	Actuate	Citizen feedback
Abdeen and Sepasgozar (2022)	City	Not applied (zero)	Community participation
Najafi et al. (2021)	City	Integrate	Community participation
Wang (2021)	City	Not applied (zero)	Transparency
Nochta et al. (2021)	City	Not applied (zero)	Socio-technical perspective
Charitonidou (2022)	City	Not applied (zero)	Universalism & socio-technical perspective

how a digital twin model could empower citizens to react to planned changes in the city and report regional issues, which can then be incorporated into a predictive decision-making process. In another study, Abdeen and Sepasgozar (2022) proposed a five-layer digital twin model to facilitate community participation in smart city planning, allowing urban infrastructure to be automatically monitored based on community reaction. The proposed digital twin model provides a promising approach to enhancing community participation in urban planning and decision-making.

- Enhancing transparency in decision-making: Digital twin models that rely on real-time data to predict and propose scenarios for influencing urban processes can lead to confusion and mistrust if citizens are not aware of the decision-making processes (Wang, 2021). Transparency helps ensure accountability and legitimacy in decision-making by allowing for review, repetition, or revision of decisions. It also opens up systems to social visibility, enabling the public to understand how decisions are made. Nochta et al. (2021) also highlight that information transparency can increase the accountability of urban managers. According to (Masoumi et al., 2023), transparency is critical in ensuring appropriate supervision. Planners and civil society can also use it to manage the impact of digital twins in social, political, and economic areas. (Sepasgozar, 2020) suggests that strategies such as transparency, digital twin modelling, and other technologies can help make digital twin technologies more benevolent.
- Enhancing evaluation of implementation before intervention: Digital twins can optimise effective decision-making processes at a city scale by enabling better insights that support better decisions (Nochta et al., 2021). Creating value by offering insights to support critical decisions about the use, maintenance, and upkeep of national and local infrastructure assets and services is the stated goal of a built environment digital twin (Nochta et al., 2021). However, access to data is still not fully available to the general public in most cases, and urban planners and managers with access to digital twins at a city scale can still modify their strategies readily (Charitonidou, 2022).

6.3.4.4.2 LIMITATIONS AND CHALLENGES OF DIGITAL TWINS IN UCPM

- Social-technical hurdles: When designing digital twin models, technical issues are often prioritised at the expense of non-material factors like social interaction, social norms, rules and regulations, cultural changes, historical values, political concerns, and associated ethics (Charitonidou, 2022). Research has highlighted the need to shift the design approach of digital twin models from a solely technical perspective to a socio-technical one (Nochta et al., 2021). Masoumi et al. (2023) suggest that digital twins should encompass broader dimensions beyond infrastructure and technology. They also criticise the idea of "data universalism" as being too narrow. Nochta et al. (2021) argue that to be more than just a technological tool, digital twin models must consider the specific characteristics of the local urban and socio-political context. The authors propose a socio-technical perspective for developing digital twin models, which includes participatory model development, reflecting the local context in the model design. The model also considers the individual and organisational learning costs associated with adopting and integrating digital twin technology into decision-making structures and processes.
- Privacy concerns: Privacy is a critical concern when developing digital twins in urban settings, as data sources are often inaccessible or difficult to produce due to factors like asset ownership, data scattering, and concerns about privacy and security. Experts in digital twin technology, urban informatics, privacy law, and human rights advocacy need to collaborate to develop a legal framework that prioritises regulations promoting transparency and responsible data usage. This framework should recognise the "right to the city" and ensure that digital twins are developed to respect privacy and uphold the fundamental human rights of people living in urban environments (Wang, 2021).
- User-friendliness: Developing user-friendly digital twin models requires specialised technical knowledge, making it challenging to create models that are accessible to a broad range of users, for example, older adult citizens. Data quality and accuracy are also crucial factors that can be challenging to gain, due to data fragmentation and lack of standardisation. Najafi et al. (2021) suggest that incorporating non-technical social, cultural, and economic considerations into digital twin models, requires sociology, anthropology, gerontology, and urban planning expertise (Najafi et al., 2021). In a later study, the authors indicate that significant investment in research and development, software development, and training is necessary to create user-friendly digital twin models, which may not be feasible for many organisations (Najafi et al., 2023).

6.3.4.5 Thread 5: Urban Modelling and Visualisation

Table 6.7 provides a summary of seven studies pertaining to digital twins in Urban Modelling and Visualisation (UM&V). The table presents an outline of the different techniques employed by the studies, such as real-time rendering, simulation of urban crowds, and tera-pixel visualisation. It showcases the various digital twin

Table 6.7 Reviewed papers in UM&V, highlighting their study scale, digital twin application level, and focus area

Study scale	*Digital Twining maturity level*	*Focus area*
City (Ssin et al., 2021)	Actuate	Visualisation
City (Huo et al., 2021)	Visualise	Real-time rendering
Building (Zhu & Wu, 2021)	Visualise	Visualisation
Individual user	Integrate	Simulate urban crowd
City (Holliman et al., 2022; Meta et al., 2021)	Visualise	Terapixel visualisation
City (Kikuchi et al., 2022)	Visualise	Future landscape/real-time rendering

maturity levels achieved at different scales and offers valuable insights into the multifaceted aspects of digital twinning within UM&V.

6.3.4.6 Exploring the benefits of digital twins in UM&V

- Future landscape visualisation: Digital twin is able to visually demonstrate the future development of urban projects in a clear and accessible manner for non-expert individuals, even those with limited technical resources. Kikuchi et al. (2022) aim to accomplish this by proposing a digital twin approach for AR in outdoor spaces. This model lets stakeholders view construction projects through on-site and off-site AR videos using their web browsers, providing a comprehensive and immersive experience.
- Diverse perspectives: Many digital twin models for UM&V are presented in a bird's-eye view, which may not accurately represent a person's direct experience of the space. Recent research aims to increase the observer's viewing angles in digital twin models. For example, Kikuchi et al. (2022) created a digital twin model for displaying AR in outdoor spaces with first-person and bird's-eye views, using precise 3D urban models. Ssin et al. (2021) propose a mixed-reality visualisation system called GeoACT, which uses a 3D virtual miniature city with real and virtual data to provide a macro-perspective view. The system also includes a 360-degree video visualisation technique for a micro-perspective view, which shows details in first-person view and supports real-time monitoring functions for city observers to monitor specific points of interest in urban space.

6.3.4.6.1 LIMITATIONS AND CHALLENGES OF DIGITAL TWINS IN UM&V

- Real-time rendering (data loading time): GIS researchers have proposed several approaches for loading and visualising urban landscapes. However, current methods for rapidly rendering in real-time require substantial memory, making it challenging to execute simulation systems outside controlled environments due to weight, power supply, and communication requirements. To address this challenge, Kikuchi et al. (2022) suggest a method for rendering AR using model-based occlusion, handling it on a personal computer. Additionally, Huo et al. (2021) propose a combined approach that uses oblique photogrammetry models to streamline real-time simulation significantly and reduce memory usage. However, these proposals are still in early development and require further investigation.

- Data operation capacity: The sheer volume of data and diverse formats in which information can be presented pose significant challenges to simulating urban layers. Current methods for launching digital twins in cities typically require excessive GPU calculations and memory usage. Recent research aims to reduce data storage and analysis capacities in digital twin models for cities. One proposed method is to employ binary encoding to improve visualisation fluency regarding data loading significantly and reduce memory usage. Another suggestion is terapixel visualisation, which creates a scalable cloud space for visualisation and supports three-dimensional tera-pixel visualisation of realistic images of urban IoT data with daily updates (Holliman et al., 2022). A proposed virtual IoT (VI) can reduce development time and costs compared to building a real IoT (RI) in a city, which requires expensive hardware installations that do not guarantee IoT functionality. Moreover, most VI datasets contain a small amount of text and numbers, allowing the proposed method to utilise even large datasets.
- Matching real and virtual images in real-time: This is a considerable challenge for digital twinning. Kikuchi et al. (2022) propose a dual digital approach that uses AR to simulate landscape visualisation in outdoor spaces. The goal is to develop a technique connecting location and sensor data from both worlds, enabling the virtual twin to continuously react to the real twin. However, video transmission between the virtual twin and its real counterpart remains challenging. It is worth noting that recent communication technologies, such as 5G, are predicted to facilitate low-latency video transmission between controllers and personal devices, which could enhance the visualisation performance of dual digital cities.

6.4 Concluding remarks, implications, and future agenda

6.4.1 Mapping the threads

After carefully examining different threads against the ANZLIC digital twin model and UDS pyramid, studies are mapped to place them in the ANZLIC model as opposed to the hierarchy graph, as shown in Figure 6.4. The occurrence of studies in each thread in the graph is explained as follows.

The studies in the UI&S thread are shown as green dots in Figure 6.4 and are mainly occurring around the integrating, energising, actuating, and automating components of the maturity hierarchy (x-axis) and the building, city, and national urban hierarchy (y-axis). At the integrated level, data from sensors and maps help city planners make better decisions about infrastructure investments, emergency management, and resource allocation using digital twins. Next, at the energise

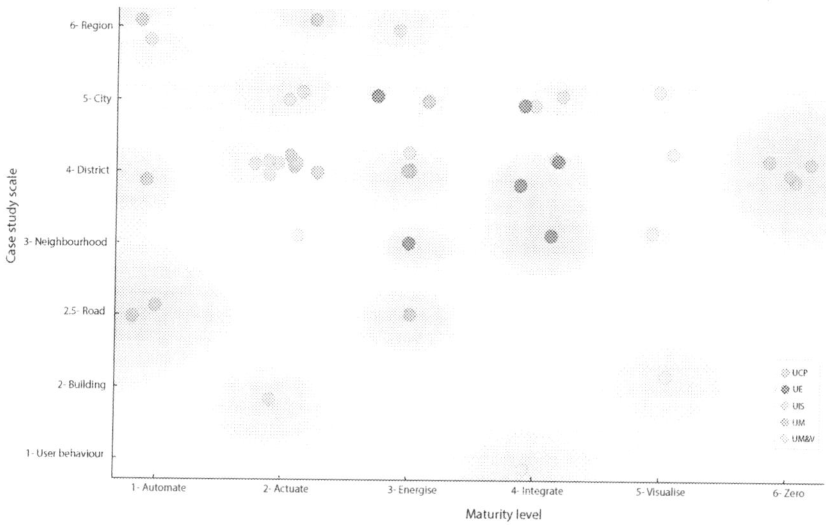

Figure 6.4 Mapping the studies in different threads in ANZLIC's maturity levels and urban decision-making levels.

level, creating a virtual model of energy systems such as power grids or street lighting can identify inefficiencies and suggest improvements to reduce energy consumption using the digital twin application. Digital twins can act as a control centre for UI&S to actuate real-time monitoring of traffic flow, signals, and energy usage in buildings and adjust heating and cooling systems to optimise energy use. To automate routine tasks and processes, freeing up staff to focus on more complex tasks, digital twins automatically generate work orders for maintenance tasks based on sensor data or dispatch emergency services in the event of natural disasters in UI&S.

Figure 6.4 – Mapping the studies in different threads in ANZLIC's maturity levels and urban decision-making levels

Further, the digital twin applications in the UI&S thread are directed towards building, city, and national levels of the hierarchy pyramid. At the building level, tasks such as optimising energy use, improving occupant comfort, and predicting potential issues can be facilitated by digital twin utilisation. Digital twins at the city level aid in modelling and simulating the entire urban environment, help city planners allocate resources in the correct directions and facilitate decision-making. Similarly, at the national level, digital twin delivers the smooth optimisation and analysis of entire urban systems and improve UI&S in an effective manner.

The studies in the UE thread are shown as red dots in Figure 6.4 and largely occur around integration and energisement in the maturity hierarchy (*x*-axis) and city (*y*-axis). By leveraging the combined and energise areas of the ANZLIC model, the digital twin applications in UE can create comprehensive and accurate models of energy consumption and generation in the urban environment and use those models to optimise energy efficiency, reduce costs, and support the transition to renewable energy sources. For the integrated area of the ANZLIC model, the key areas are the provision of data from energy metres, building management systems, weather forecasts, and other sources to create a comprehensive model of energy consumption and generation in the urban environment. On the other hand, the energised area of the ANZLIC model involves using the digital twin to simulate and optimise energy systems in real time. It involves using better energy distribution across the city, identifying opportunities to reduce energy consumption, and further improving energy systems. In summary, the digital twin applications in UE revolve around the city level of the UDS pyramids because they provide a comprehensive and detailed view of the city's energy consumption and generation, which is critical for informed decisions at the city level.

The studies in the UM thread are shown as orange dots in Figure 6.4 and mainly occur around energise, actuate, and automate on the maturity hierarchy (x-axis) and road, city and national on the urbanisation hierarchy (y-axis). Digital twins can energise the UM systems, such as public transportation or electric vehicle charging stations. By creating a virtual model of the system and analysing data from sensors, the digital twin can identify inefficiencies and suggest improvements to reduce energy consumption. At the actuate level, digital twins can act as a control centre for UM systems by integrating real-time data from various sources, such as traffic sensors or public transportation schedules, to make automated decisions to optimise system performance. For instance, the digital twin can reroute buses to avoid traffic congestion or adjust traffic signal timings to reduce commuter wait times. Finally, as UM systems are becoming increasingly automated, with the introduction of autonomous vehicles and smart traffic management systems, digital twins can help develop and test these automated systems by creating a virtual environment to simulate and optimise their performance.

Further, digital twin applications in the UM are primarily focused on the road, city, and national levels of the UDS pyramid. At the road level, digital twins can model and simulate traffic flow and pedestrian movements by identifying and predicting potential bottlenecks or safety hazards and optimising traffic flow in real time. Similarly, at a city level, the modelling of public transportation, roads, and parking helps city planners make data-driven decisions about where to invest in new infrastructure, how to optimise routes, and how to improve overall mobility in the city. Lastly, at the national level, digital twins can be used to analyse and optimise transportation systems nationwide. This includes interstate highways and national rail networks. By creating virtual replicas of the systems, policymakers can analyse the impact of different policies, such as investment in high-speed rail or road tolls, and make informed decisions about improving national mobility.

The studies in the UCPM thread are shown as blue dots in Figure 6.4 and mainly occur around those located on the maturity hierarchy (x-axis) and at the city and district level on the urbanisation hierarchy (y-axis). As the locate layer provides the foundation for all other layers, the information is essential for creating an accurate digital twin that can be used to model and simulate the community's behaviour (Dani et al., 2023). Also, as the locate layer provides a common reference point to other layers, stakeholders can use the identical digital twin to collaborate in decisions about governance and leading the community (Wang & Wang, 2021). For instance, city planners can use locate to model the impact of new development projects on the community. In contrast, residents can use the exact digital twin to provide feedback and suggest improvement. There are various reasons why the UCPM thread is lacking in other layers of the ANZLIC model. Firstly, creating a digital twin can be complex and resource-intensive and requires accurate and up-to-date data, which is hard to obtain, especially for some of the thematic areas of UCPM, such as social and economic data. Secondly, as funding is limited for UCPM, creating a technology and infrastructure-rich digital twin model can be costly. Finally, the lack of potential benefits of a digital twin is limited in the areas of UCPM, and stakeholders are sceptical about honing the actual benefits. However, one study in this thread by White et al. (2021) comes under the actuating area where smart sensors are used to collect real-time information on buildings. Another example of UCPM in the actuating layer consists of optimising data for traffic flow in the community by using sensors and cameras on roads and intersections.

In addition, the studies in this thread are primarily at the city and district levels of the hierarchy pyramid, the location of the most pressing challenges of urban management, such as traffic congestion, air pollution, and access to services like water and sanitation. Further, factors such as the location of many stakeholders in the city and associated districts and regions, the presence of urban infrastructure systems such as transport networks, water and energy systems, and public spaces, and the availability of a broad network for creating digital twin at city and district level, corroborate with the focus of UCPM studies in these areas.

Studies in the UM&V thread are shown as yellow dots in Figure 6.4 and mainly occur around visualise (x-axis) and at all levels of the hierarchy pyramid scale (y-axis). Two studies lie under the actuate and integrate layers, respectively. Also, the studies are spread throughout the hierarchy pyramid on the x-axis, reflecting the significance of UM&V across different areas. The visualisation component of the ANZLIC digital twin model is critical in UM&V because it enables stakeholders to interact with the digital twin and gain a better understanding of the physical environment. By creating realistic images of the city and its components, digital twin applications can help stakeholders visualise different scenarios and make more informed decisions. In terms of integrated components, digital twins can be used to integrate data from various sources, such as sensors, GIS data, and other real-time traffic data, to create a complete and more accurate city model. This integrated data can then be used to simulate and analyse different scenarios, such as the impact of new developments on traffic flow or the effect of weather events on infrastructure. Digital twins can also be integrated with virtual and augmented realities to create immersive and interactive visualisations of the city to help better understand the physical environment. Lastly, digital twins can act in response to real-time data and simulations in the actuating element of the maturity hierarchy. For instance, a digital twin of a city's traffic system can use traffic data to adjust traffic light timings and optimise traffic flow. Similarly, it can be used for energy usage to modify mechanical systems and optimise energy efficiency.

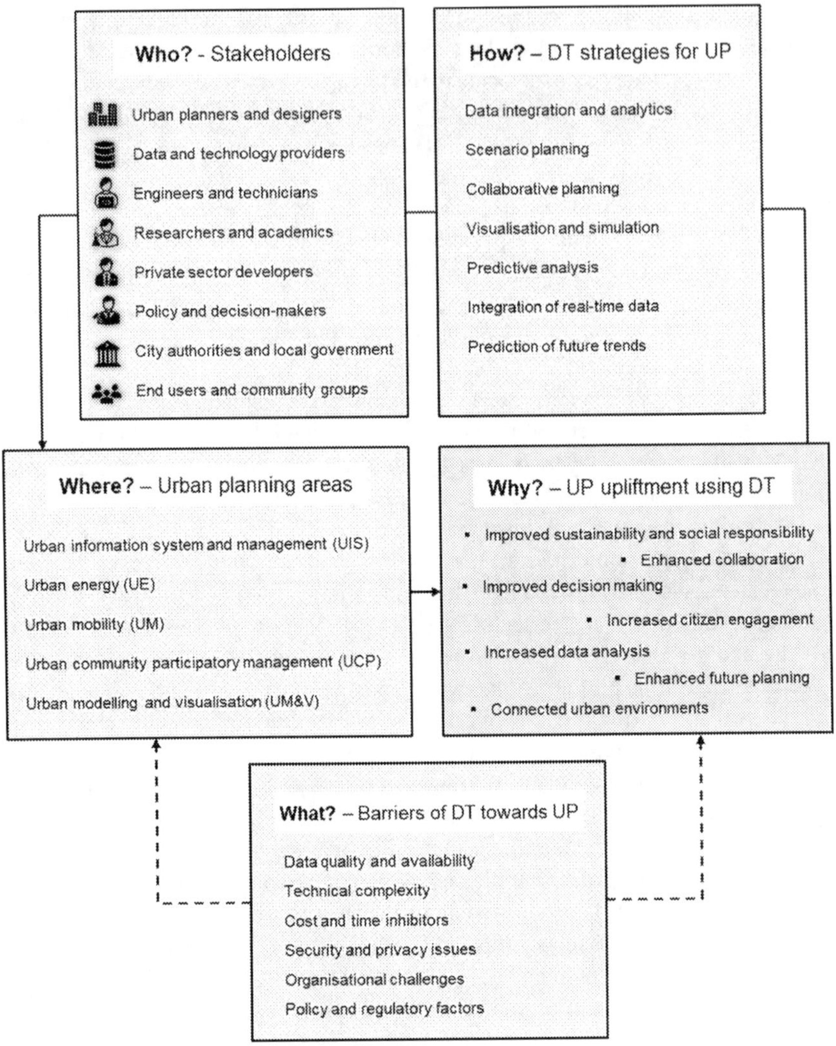

Figure 6.5 The W4-H1 framework for digital twins in planning decision support systems.

6.4.2 *A digital twin-enabled urban decision-making framework*

To aid future scholarly work, academics, practitioners, and other stakeholders of the topic, this study developed a framework to cover the boundaries of digital twin and urban planning integration (see Figure 6.5). A comprehensive literature review and content analysis of the prior literature have unravelled various research gaps and opportunities. The framework, named W4-H1, lays out the roles of different stakeholders (the who), the digital twin strategies for urban planning (the how), the relevant areas of urban planning (the where), the potential barriers raised by the digital twin in urban planning (what), and finally, the uplift of urban planning using digital twins (why) of the framework (Caprari et al., 2022).

Table 6.8 Knowledge gaps and future research areas in each thread studied in this chapter

Thread	*Knowledge gaps*	*Future research*
Urban Information Systems and Management (UI&S)	• Integration with real-time data (Eri & Elnæs, 2023) – data from sources such as IoT sensors, satellite imagery, and social media platforms. • Privacy and security – a collection of data without theft or breach issues. • Interoperability and standardisation – caused by the involvement of multiple technologies and platforms. • Human-centric design – impact on human well-being and social equity is unknown. • Scalability – its resource-intensive nature makes it challenging to scale up to larger urban areas.	• Developing efficient data management and analytics techniques to support real-time decision-making. • Use technologies such as Blockchain to address security and privacy issues while maintaining the benefits of digital twins. • Developing standards and guidelines to ensure compatibility and ease of integration. • More inclusive and participatory urban design using digital twins • More efficient algorithms and techniques for digital twin implementation to support scalability.
Urban Energy (UE)	• Integration of renewable energy sources – lack of optimisation of solar and wind power using digital twin models. • Energy storage – lack of storage causing intermittency and grid stability issues. • Demand response – lack of demand response to adjust consumption during peak hours using a digital twin. • Smart grid management – lack of streamlined digital twin models to optimise the operation of smart grids to improve efficiency and reliability of energy distribution. • Human-centric design – lack of consideration for the impact of human behaviour and well-being in the urban energy systems • Multi-domain modelling – lack of application of multi-domains such as energy generation, distribution, and consumption in urban energy digital twins.	• Developing more accurate models for predicting production, consumption and optimisation in urban environments. • more practical use of digital twins to optimise the design and operation of energy storage systems in urban areas. • To predict and optimise the impact of demand response programs in urban energy systems. • Focused digital twin models to understand and predict grid stability, resilience, demand and supply, and operations optimisation. • Future research should focus on using digital twins to support user-centric and participatory urban energy systems design. • Need to capture complex interactions between these domains and support optimisation of urban energy systems.

(*Continued*)

Table 6.8 (Continued)

Thread	*Knowledge gaps*	*Future research*
Urban Mobility (UM)	• Data availability – challenging data availability due to a lack of standardised data formats and limited access from different sources. • Model accuracy doubted – due to physical systems' complexity, data input uncertainties, and limitations with modelling techniques. • Interdisciplinary integration – between various fields, such as urban planning, energy engineering, computer science, and data analytics.	• Connected and autonomous vehicles (CAVs) – integration with CAVs for predicting traffic flow, reducing congestion, and enhancing safety. • Integrated mobility systems – integration with digital twin model for better public transit, bike sharing, and car sharing to create a seamless mobility system. • Smart traffic management – optimises real-time traffic flow traffic signals and coordinates CAVs. • Pedestrian and bicycle infrastructure – analysing walkability and bike-ability improves safety and reduces conflict with other transport modes. • Mobility as a service (MaaS) – optimise MaaS platforms to provide integrated transport services to the users.
Urban Community Participatory Management (UCP)	• Citizen engagement – lack of engaging diverse groups of citizens due to low access to digital tools. • Data privacy and security – need to develop effective methods to protect personal data while enabling effective participation. • Interdisciplinary integration – challenges in integrating different disciplines due to low levels of communication and collaboration.	• Collaborative urban planning – facilitation of collaborative urban planning involving citizens, community groups, and stakeholders. • Community resilience – use of digital twins to simulate the resilience of urban communities to different shocks and stressors such as natural disasters, pandemics, and climate change. • Data visualisation and communication use digital twins to develop effective data visualisation and communication, using virtual reality and gamification tools for users to understand complex urban systems. • Impact assessment – use of digital twins to assess the impact of different policies and interventions on the community, allowing citizens and decision-makers to evaluate the effectiveness of different approaches.

(*Continued*)

Table 6.8 (Continued)

Thread	*Knowledge gaps*	*Future research*
Urban Modelling and Visualisation (UM&V)	• Availability and integration of data – ineffective availability and integration due to lack of standardised data formats and access from different sources. • Accuracy and effectiveness of model – lack of accuracy due to physical systems' complexity, data inputuncertainties, and limitations with modelling techniques. • Integration between different disciplines – a challenge to integrate various disciplines requiring effective communication and collaboration.	• Real-time updates – opportunities for updating the digital twin model in real-time for up-to-date planning. • Virtual and augmented reality – enabling stakeholders to interact with the urban environment using the likes of virtual and augmented realities. • Simulation and scenario testing – opportunities for more informed decision-making by simulating different scenarios and their potential impacts. • Predictive analysis – development opportunities for predictive models to enable stakeholders to anticipate and plan • Spatial data analysis – opportunities to analyse spatial data, including 3D models and GIS, to identify patterns and trends in urban environments.

This study frames the fundamentals of integrative interfaces between digital twins and urban planning using the W4-H1 framework. The "who" factor refers to the importance of stakeholder integration to realise the collaborative benefits of digital twins in urban planning. As all stakeholders shown in the framework are equally essential to facilitate the transition of a digital twin, the government plays a significant role in the unification and transformation of this sector. The "how" factor of the framework refers to the critical areas and strategies where digital twins can substantially enhance their value in urban planning systems. Strategies such as "data integration and analytics," "scenario and collaborative planning," "visualisation and simulation," "predictive analysis," and "prediction of future trends" are significant focuses where digital twins can unravel the true potential of urban planning. The "where" factor of the framework refers to the threads identified in this study where vast opportunities for digital twins can be used to ensure improvements and relevance in the context of urban planning. The "what" factor of the framework provides inhibiting elements when transitioning digital twin applications in the urban planning domain. As such, efficient and effective digital twin models are required to eliminate the detrimental effects of these inhibitors, with the assistance of proper managerial governance and other driving factors. Lastly, the "why" factor of the framework strengthens the need for digital twin implementation in urban planning to achieve "improved sustainability and social responsibility," "enhanced collaboration," "improved decision making," "increased citizen engagement," "increased data analysis," "enhanced future planning," and "connected urban environments." Doing so will not only provide economic, social, and environmental gains from urban planning but also achieve technical, stakeholder, and technological benefits.

6.4.3 Future research agenda

A comprehensive literature review and content analysis of the prior literature have unravelled various research gaps and opportunities. The knowledge gaps were identified, and future research was proposed that was implicitly mentioned by previous research work. Table 6.8 summarises the relevant themes with knowledge gaps and prospects for filling them.

Ultimately, it is imperative to acknowledge the inherent limitations of relying solely on a literature review and the need for more diverse sources of data and perspectives in future studies. The theoretical nature of many of the papers reviewed, coupled with the high costs and specialised technical expertise required for implementing digital twin technology, means that there is limited empirical evidence on its effectiveness in improving urban management and decision-making processes. Additionally, ethical and legal considerations may have limited the disclosure of full case studies, limiting our understanding of the technology's potential impact. This study serves as a foundation for future empirical investigations, encouraging scholars to delve into the practical applications and impacts of digital twin technology on urban environments.

Appendix A

Table A.1 Selected articles for peer review

References	*Publisher title*	*Journal title*	*Context*
Abdeen and Sepasgozar (2022)	MDPI	Environmental Sciences Proceedings	Australia
Akroyd et al. (2022)	Cambridge Core	Data-Centric Engineering	United Kingdom
Anda et al. (2021)	Elsevier	Transportation Research Part C: Emerging Technologies	Singapore
Argota Sánchez-Vaquerizo (2021)	MDPI	Computational Social Science	Spain
Badawi et al. (2021)	MDPI	Sensors	Canada

(*Continued*)

Table A.1 (Continued)

References	*Publisher title*	*Journal title*	*Context*
Bartos and Kerkez (2021)	Elsevier	Environmental Modelling & Software	USA
Broekman et al. (2021)	Elsevier	Transportation Engineering	South Africa
Buckley et al. (2021)	MDPI	Energies	Ireland
Bujari et al. (2021)	MDPI	Sensors	Italy and USA
Charissis et al. (2021)	MDPI	Applied Sciences	USA
Charitonidou (2022)	SAGE	International Journal of Architectural Computing	Greece
Ghandar et al. (2021)	IEEE	Access	China
Gürdür Broo et al. (2022)	Elsevier	Automation in Construction	United Kingdom
Gutierrez-Franco et al. (2021)	MDPI	Sustainability	USA
HosseiniHaghighi et al. (2022a)	Elsevier	Energy and Buildings	Canada
HosseiniHaghighi et al. (2022b)	Elsevier	Sustainable Cities and Society	Canada
Holliman et al. (2022)	IEEE	Transactions on Cloud Computing	United Kingdom
Hu et al. (2022)	IEEE	Transactions on Industrial Informatics	China
Huo et al. (2021)	MDPI	International Journal of Geo-Information	China
Jiang et al. (2022)	Elsevier	Sustainable Cities and Society	United Kingdom
Kikuchi et al. (2022)	Oxford Academic	Journal of Computational Design and Engineering	United Kingdom
Lee et al. (2022a)	MDPI	Remote Sensing	Korea
Lenfers et al. (2021)	MDPI	Sustainability	Germany
Leung et al. (2022)	Elsevier	International Journal of Production Economics	United Kingdom
Li et al. (2022)	Elsevier	Future Generation Computer Systems	China
Major et al. (2021)	IEEE	Instrumentation & Measurement Magazine	Norway
Marai et al. (2021)	IEEE	Network	Finland
Marchione and Ruperto (2022)	WIT Press	International Journal of Energy Production and Management	Italy
Meta et al. (2021)	Hindawi	Complexity	Spain
Najafi et al. (2021)	IEEE	Intelligent Environments	The Netherlands
Nochta et al. (2021)	Tylor & Francis	Journal of Urban Technology	United Kingdom

(*Continued*)

Table A.1 (Continued)

References	*Publisher title*	*Journal title*	*Context*
Orozco-Messana et al. (2021)	MDPI	Sustainability	Spain
Pang et al. (2020)	IEEE	Tsinghua Science and Technology	China
Pedersen et al. (2021)	MDPI	Water	Denmark
Raes et al. (2022)	IEEE	Internet Computing	Greece, Czech Republic, Belgium
Saroj et al. (2021)	ASCE	Journal of Transportation Engineering, Part A: Systems	USA
Ssin et al. (2021)	IxD&A	Interaction Design & Architecture(s) Journal	Korea
Steinmetz et al. (2022)	IEEE	Transactions on Emerging Topics in Computing	Germany
Truu et al. (2021)	MDPI	Water	Estonia and Sweden
Van de Vyvere and Colpaert (2022)	Springer	European Transport Research Review	Belgium
Wang (2021)	Elsevier	Expert Systems with Applications	China
White et al. (2021)	Elsevier	Cities	Ireland
Zhao et al. (2022)	Frontiers	Frontiers in Plant Science	China
Zhu and Wu (2021)	MDPI	Remote Sensing	Australia

References

Abdeen, F. N., & Sepasgozar, S. M. E. (2022). City digital twin concepts: A vision for community participation. *The 3rd Built Environment Research Forum, 19*. https://doi.org/10.3390/environsciproc2021012019.

Akroyd, J., Harper, Z., Soutar, D., Farazi, F., Bhave, A., Mosbach, S., & Kraft, M. (2022). Universal digital twin: Land use. *Data-Centric Engineering, 3*, e3. https://doi.org/10.1017/dce.2021.21.

Anda, C., Ordonez Medina, S. A., & Axhausen, K. W. (2021). Synthesising digital twin travellers: Individual travel demand from aggregated mobile phone data. *Transportation Research Part C: Emerging Technologies, 128*, 103118. https://doi.org/10.1016/j.trc.2021.103118.

ANZLIC. (2019). *Principles for Spatially Enabled Digital Twins of the Built and Natural Environment in Australia*. ANZLIC Barton.

Argota Sánchez-Vaquerizo, J. (2021). Getting real: The challenge of building and validating a large-scale digital twin of Barcelona's traffic with empirical data. *ISPRS International Journal of Geo-Information, 11*(1), 24. https://doi.org/10.3390/ijgi11010024.

Badawi, H. F., Laamarti, F., & El Saddik, A. (2021). Devising digital twins DNA paradigm for modeling ISO-based city services. *Sensors, 21*(4), 1047. https://doi.org/10.3390/s21041047.

Bartos, M., & Kerkez, B. (2021). Pipedream: An interactive digital twin model for natural and urban drainage systems. *Environmental Modelling & Software, 144*, 105120. https://doi.org/10.1016/j.envsoft.2021.105120.

Broekman, A., Gräbe, P. J., & Steyn, W. J. v. (2021). Real-time traffic quantization using a mini edge artificial intelligence platform. *Transportation Engineering, 4*, 100068. https://doi.org/10.1016/j.treng.2021.100068.

Buckley, N., Mills, G., Letellier-Duchesne, S., & Benis, K. (2021). Designing an energy-resilient neighbourhood using an urban building energy model. *Energies, 14*(15), 4445. https://doi.org/10.3390/en14154445.

Bujari, A., Calvio, A., Foschini, L., Sabbioni, A., & Corradi, A. (2021). A digital twin decision support system for the urban facility management process. *Sensors, 21*(24), 8460. https://doi.org/10.3390/s21248460.

Caprari, G., Castelli, G., Montuori, M., Camardelli, M., & Malvezzi, R. (2022). Digital twin for urban planning in the green deal era: A state of the art and future perspectives. *Sustainability, 14*(10), Article 10. https://doi.org/10.3390/su14106263.

Charissis, V., Falah, J., Lagoo, R., Alfalah, S. F. M., Khan, S., Wang, S., … Drikakis, D. (2021). Employing emerging technologies to develop and evaluate in-vehicle intelligent systems for driver support: Infotainment AR HUD case study. *Applied Sciences, 11*(4), 1397. https://doi.org/10.3390/app11041397.

Charitonidou, M. (2022). Urban scale digital twins in data-driven society: Challenging digital universalism in urban planning decision-making. *International Journal of Architectural Computing, 20*(2), 238–253. https://doi.org/10.1177/14780771211070005.

Dani, A. A. H., Supangkat, S. H., Lubis, F. F., Nugraha, I. G. B. B., Kinanda, R., & Rizkia, I. (2023). Development of a smart city platform based on digital twin technology for monitoring and supporting decision-making. *Sustainability, 15*(18), Article 18. https://doi.org/10.3390/su151814002.

Deng, T., Zhang, K., & Shen, Z.-J. (Max). (2021). A systematic review of a digital twin city: A new pattern of urban governance toward smart cities. *Journal of Management Science and Engineering, 6*(2), 125–134. https://doi.org/10.1016/j.jmse.2021.03.003.

Eri, A. B., & Elnæs, I. A. (2023). *The Emergence of Digital Twin Technology in Urban Planning: A Study of Perceptions, Opportunities and Barriers* [Master thesis, Norwegian University of Life Sciences]. https://nmbu.brage.unit.no/nmbu-xmlui/handle/11250/3078973.

Ghandar, A., Ahmed, A., Zulfiqar, S., Hua, Z., Hanai, M., & Theodoropoulos, G. (2021). A decision support system for urban agriculture using digital twin: A case study with aquaponics. *IEEE Access, 9*, 35691–35708. https://doi.org/10.1109/ACCESS.2021.3061722.

Gürdür Broo, D., Bravo-Haro, M., & Schooling, J. (2022). Design and implementation of a smart infrastructure digital twin. *Automation in Construction, 136*, 104171. https://doi.org/10.1016/j.autcon.2022.104171.

Gutierrez-Franco, E., Mejia-Argueta, C., & Rabelo, L. (2021). Data-driven methodology to support long-lasting logistics and decision making for urban last-mile operations. *Sustainability, 13*(11), 6230. https://doi.org/10.3390/su13116230.

Holliman, N. S., Antony, M., Charlton, J., Dowsland, S., James, P., & Turner, M. (2022). Petascale cloud supercomputing for terapixel visualization of a digital twin. *IEEE Transactions on Cloud Computing, 10*(1), 583–594. https://doi.org/10.1109/TCC.2019.2958087.

HosseiniHaghighi, S., de Uribarri, P. M. Á., Padsala, R., & Eicker, U. (2022a). Characterizing and structuring urban GIS data for housing stock energy modelling and retrofitting. *Energy and Buildings, 256*, 111706. https://doi.org/10.1016/j.enbuild.2021.111706.

HosseiniHaghighi, S., Panchabikesan, K., Dabirian, S., Webster, J., Ouf, M., & Eicker, U. (2022b). Discovering, processing and consolidating housing stock and smart thermostat data in support of energy end-use mapping and housing retrofit program planning. *Sustainable Cities and Society, 78*, 103640. https://doi.org/10.1016/j.scs.2021.103640.

Hu, C., Fan, W., Zeng, E., Hang, Z., Wang, F., Qi, L., & Bhuiyan, M. Z. A. (2022). Digital twin-assisted real-time traffic data prediction method for 5G-enabled internet of vehicles. *IEEE Transactions on Industrial Informatics, 18*(4), 2811–2819. https://doi.org/10.1109/TII.2021.3083596.

Huo, Y., Yang, A., Jia, Q., Chen, Y., He, B., & Li, J. (2021). Efficient visualization of large-scale oblique photogrammetry models in unreal engine. *ISPRS International Journal of Geo-Information, 10*(10), 643. https://doi.org/10.3390/ijgi10100643.

Jiang, F., Ma, L., Broyd, T., Chen, W., & Luo, H. (2022). Digital twin enabled sustainable urban road planning. *Sustainable Cities and Society, 78*, 103645. https://doi.org/10.1016/j.scs.2021.103645.

Ketzler, B., Naserentin, V., Latino, F., Zangelidis, C., Thuvander, L., & Logg, A. (2020). Digital twins for cities: A state of the art review. *Built Environment, 46*(4), 547–573. https://doi.org/10.2148/benv.46.4.547.

Kikuchi, N., Fukuda, T., & Yabuki, N. (2022). Future landscape visualization using a city digital twin: Integration of augmented reality and drones with implementation of 3D model-based occlusion handling. *Journal of Computational Design and Engineering, 9*(2), 837–856. https://doi.org/10.1093/jcde/qwac032.

Lee, A., Lee, K.-W., Kim, K.-H., & Shin, S.-W. (2022a). A geospatial platform to manage large-scale individual mobility for an urban digital twin platform. *Remote Sensing, 14*(3), 723. https://doi.org/10.3390/rs14030723.

Lee, Y. L., Yeow, C. Y., Lim, M. H., Woon, K. S., & Woon, Y. B. (2022b). Preliminary study on the rapid assembly emergency shelter. In *E3S Web of Conferences*.

Lenfers, U. A., Ahmady-Moghaddam, N., Glake, D., Ocker, F., Osterholz, D., Ströbele, J., & Clemen, T. (2021). Improving model predictions—Integration of real-time sensor data into a running simulation of an agent-based model. *Sustainability, 13*(13), 7000. https://doi.org/10.3390/su13137000.

Leung, E. K. H., Lee, C. K. H., & Ouyang, Z. (2022). From traditional warehouses to Physical Internet hubs: A digital twin-based inbound synchronization framework for PI-order management. *International Journal of Production Economics, 244*, 108353. https://doi.org/10.1016/j.ijpe.2021.108353.

Li, X., Liu, H., Wang, W., Zheng, Y., Lv, H., & Lv, Z. (2022). Big data analysis of the Internet of Things in the digital twins of smart city based on deep learning. *Future Generation Computer Systems, 128*, 167–177. https://doi.org/10.1016/j.future.2021.10.006.

Major, P., Li, G., Hildre, H. P., & Zhang, H. (2021). The use of a data-driven digital twin of a smart city: A case study of Ålesund, Norway. *IEEE Instrumentation & Measurement Magazine, 24*(7), 39–49. https://doi.org/10.1109/MIM.2021.9549127.

Marai, O. E., Taleb, T., & Song, J. (2021). Roads infrastructure digital twin: A step toward smarter cities realization. *IEEE Network, 35*(2), 136–143. https://doi.org/10.1109/MNET.011.2000398.

Marchione, P., & Ruperto, F. (2022). Prototyping a digital twin – A case study of a 'U-shaped' military building. *International Journal of Energy Production and Management, 5*(4), 83–94. https://doi.org/10.2495/EQ-V7-N1-83-94.

Masoumi, H., Shirowzhan, S., Eskandarpour, P., & Pettit, C. J. (2023). City digital twins: Their maturity level and differentiation from 3D city models. *Big Earth Data, 7*(1), 1–36. https://doi.org/10.1080/20964471.2022.2160156.

Meta, I., Serra-Burriel, F., Carrasco-Jiménez, J. C., Cucchietti, F. M., Diví-Cuesta, C., García Calatrava, C., … Eguskiza Martínez, I. (2021). The Camp Nou Stadium as a testbed for city physiology: A modular framework for urban digital twins. *Complexity, 2021*, 1–15. https://doi.org/10.1155/2021/9731180.

Moher, D., Liberati, A., Tetzlaff, J., Altman, D. G., & Group, T. P. (2009). Preferred reporting items for systematic reviews and meta-analyses: The PRISMA statement. *PLOS Medicine, 6*(7), e1000097. https://doi.org/10.1371/journal.pmed.1000097.

Najafi, P., Mohammadi, M., Le Blanc, P. M., & Van Wesemael, P. (2021). Experimenting a healthy ageing community in immersive virtual reality environment: The case of world's longest-lived populations. In *2021 17th International Conference on Intelligent Environments (IE)* (pp. 1–5). https://doi.org/10.1109/IE51775.2021.9486595.

Najafi, P., Mohammadi, M., Le Blanc, P. M., & van Wesemael, P. (2022). Insights into placemaking, senior people, and digital technology: A systematic quantitative review. *Journal of Urbanism: International Research on Placemaking and Urban Sustainability*, 1–30. https://doi.org/10.1080/17549175.2022.2076721.

Najafi, P., Mohammadi, M., van Wesemael, P., & Le Blanc, P. M. (2023). A user-centred virtual city information model for inclusive community design: State-of-art. *Cities, 134*, 104203. https://doi.org/10.1016/j.cities.2023.104203.

Nochta, T., Wan, L., Schooling, J. M., & Parlikad, A. K. (2021). A socio-technical perspective on urban analytics: The case of city-scale digital twins. *Journal of Urban Technology, 28*(1–2), 263–287. https://doi.org/10.1080/10630732.2020.1798177.

Orozco-Messana, J., Iborra-Lucas, M., & Calabuig-Moreno, R. (2021). Neighbourhood modelling for urban sustainability assessment. *Sustainability, 13*(9), 4654. https://doi.org/10.3390/su13094654.

Pang, J., Li, J., Xie, Z., Huang, Y., & Cai, Z. (2020). *Collaborative City Digital Twin for Covid-19 Pandemic: A Federated Learning Solution* (arXiv:2011.02883). arXiv. http://arxiv.org/abs/2011.02883.

Pedersen, A. N., Borup, M., Brink-Kjær, A., Christiansen, L. E., & Mikkelsen, P. S. (2021). Living and prototyping digital twins for urban water systems: Towards multi-purpose value creation using models and sensors. *Water, 13*(5), 592. https://doi.org/10.3390/w13050592.

Raes, L., Michiels, P., Adolphi, T., Tampere, C., Dalianis, A., McAleer, S., & Kogut, P. (2022). DUET: A framework for building interoperable and trusted digital twins of smart cities. *IEEE Internet Computing, 26*(3), 43–50. https://doi.org/10.1109/MIC.2021.3060962.

Saroj, A. J., Roy, S., Guin, A., & Hunter, M. (2021). Development of a connected corridor real-time data-driven traffic digital twin simulation model. *Journal of Transportation Engineering, Part A: Systems, 147*(12), 04021096. https://doi.org/10.1061/JTEPBS.0000599.

Sepasgozar, S. M. E. (2020). Digital twin and cities. In *The Palgrave Encyclopedia of Urban and Regional Futures* (pp. 1–6). Springer International Publishing. https://doi.org/10.1007/978-3-030-51812-7_253-1.

Sepasgozar, S. M. E. (2021). Differentiating digital twin from digital shadow: Elucidating a paradigm shift to expedite a smart, sustainable built environment. *Buildings, 11*(4), Article 4. https://doi.org/10.3390/buildings11040151.

Srinivas, H. (2015). *The Decision-Making Pyramid* (E-003; GDRC Reseaarch Output). Global Development Research Center. https://www.gdrc.org/decision/pyramid.html.

Ssin, S., Cho, H., & Woo, W. (2021). GeoACT: Augmented control tower using virtual and real geospatial data. *Interaction Design and Architecture(s), 48*, 122–142. https://doi.org/10.55612/s-5002-048-006.

Steinmetz, C., Schroeder, G. N., Rodrigues, R. N., Rettberg, A., & Pereira, C. E. (2022). Key-components for digital twin modeling with granularity: Use case car-as-a-service. *IEEE Transactions on Emerging Topics in Computing, 10*(1), 23–33. https://doi.org/10.1109/TETC.2021.3131532.

Truu, M., Annus, I., Roosimägi, J., Kändler, N., Vassiljev, A., & Kaur, K. (2021). Integrated decision support system for pluvial flood-resilient spatial planning in urban areas. *Water, 13*(23), 3340. https://doi.org/10.3390/w13233340.

Vaismoradi, M., Jones, J., Turunen, H., & Snelgrove, S. (2016). Theme development in qualitative content analysis and thematic analysis. *Journal of Nursing Education and Practice*. https://core.ac.uk/download/pdf/225904998.pdf

Van de Vyvere, B., & Colpaert, P. (2022). Using ANPR data to create an anonymized linked open dataset on urban bustle. *European Transport Research Review, 14*(1), 17. https://doi.org/10.1186/s12544-022-00538-1.

Wang, B. (2021). The seductive smart city and the benevolent role of transparency. *Interaction Design and Architecture(s), 48*, 100–121. https://doi.org/10.55612/s-5002-048-005.

Wang, B. T., & Wang, C. M. (Eds.). (2021). *Automating Cities: Design, Construction, Operation and Future Impact*. Springer. https://doi.org/10.1007/978-981-15-8670-5.

Wang, H., Chen, X., Jia, F., & Cheng, X. (2023). Digital twin-supported smart city: Status, challenges and future research directions. *Expert Systems with Applications, 217*, 119531. https://doi.org/https://doi.org/10.1016/j.eswa.2023.119531.

White, G., Zink, A., Codecá, L., & Clarke, S. (2021). A digital twin smart city for citizen feedback. *Cities, 110*, 103064. https://doi.org/10.1016/j.cities.2020.103064.

Wu, H., Ji, P., Ma, H., & Xing, L. (2023). A comprehensive review of digital twin from the perspective of total process: Data, models, networks and applications. *Sensors, 23*(19), Article 19. https://doi.org/10.3390/s23198306.

Ye, X., Du, J., Han, Y., Newman, G., Retchless, D., Zou, L., Ham, Y., & Cai, Z. (2022). Developing human-centered urban digital twins for community infrastructure resilience: A research agenda. *Journal of Planning Literature*, 08854122221137861. https://doi.org/10.1177/08854122221137861.

Zhao, D., Li, X., Wang, X., Shen, X., & Gao, W. (2022). Applying digital twins to research the relationship between urban expansion and vegetation coverage: A case study of natural preserve. *Frontiers in Plant Science, 13*, 840471. https://doi.org/10.3389/fpls.2022.840471.

Zhu, J., & Wu, P. (2021). Towards effective BIM/GIS data integration for smart city by integrating computer graphics technique. *Remote Sensing, 13*(10), 1889. https://doi.org/10.3390/rs13101889.

7 Web-based BIM-GIS integration workflows for making cities' digital twin 3D models

Sara Shirowzhan, Subin Mecheril Binoy and Samad M. E. Sepasgozar

City digital twin (CDT) is a trending and emerging direction of technology for cities' visualisation, simulation and analytics. This is becoming more important for further developing smart cities. However, many 3D models of CDT created in 3D web scenes lack the details of the models built, known as BIM models and also lack the capabilities of GIS software for deriving insights. Some studies suggest BIM-GIS integration for benefitting from detailed BIM models and analytics within a GIS environment. However, there is a lack of research comparing the BIM-GIS integration workflows for exploring the issues of interoperability and finding the best workflow for shadow analysis, which is one of the most important planning considerations. Indeed, the issue with many three-dimensional (3D) models of cities, mistakenly known as digital twins of cities, is that the 3D models of the buildings only extrude building footprints according to the height of the buildings. The extruded building footprints do not represent the buildings' specifications, such as the materials, sizes of openings, roof type, and materials, and a building's interior information, such as the room spaces, wall thicknesses, and floor layouts. This study fills the gap by comparative analysis of the workflows of BIM-GIS integration for applications in digital twinning cities, shadow analysis and the least interoperability issue resulting in data loss in the process of the integration. Our results show that the workflows using open-source platforms provide seamless integration of BIM and GIS and provide better visual outcomes for shadow analysis. This chapter also presents details of the data types and formats that are valuable to scholars and practitioners in the field to encourage workflows and details to be replicated in different contexts. The value of this study lies in introducing the best approach to BIM-GIS integration for better data-driven decision-making in cities, leading to smarter cities.

7.1 Introduction

In the last few years, GIS systems have been used for modelling, analysing, and visualising urban areas and details of the built environment (Gawley and McKenzie 2022; Debnath, Pettit et al. 2023; Xu, Shirowzhan et al. 2023). In GIS, all the information is georeferenced to provide an abstract model of reality. There is a

DOI: 10.1201/9781003507000-7

significant interest in the development of the representation of more realistic urban features, which is leading to the development of 3D GIS (Badwi, Ellaithy et al. 2022). A 3D GIS system provides a more realistic representation of real-world phenomena and allows advanced spatial analysis; however, due to the complexity of surface and subsurface geometries, there are many complications in the development of integrated 3D solutions (Abdul-Rahman and Pilouk 2008; Wang and Li 2022; Abdeen, Shirowzhan et al. 2023).

Building information modelling (BIM) is an information-rich and model-centric process that enables architects, engineers, and planners to create a multi-dimensional model of urban features such as buildings and infrastructure elements for various projects, including construction projects (Sepasgozar, Khan et al. 2023). BIM can increase the clarity of project dimensions and intentions for all stakeholders and ensure data fidelity across the asset utilisation period owing to the highly detailed and precise modelling techniques employed by BIM (Lee, Salama et al. 2014; Sepasgozar, Costin et al. 2022). In a construction process, either from planning or execution to the maintenance of the projects, it is necessary to have seamless information sharing with all the stakeholders involved (Manolova 2018). Building digital models is not usually integrated with the surrounding environment. For planning a new building, the model's information is insufficient to analyse various urban design parameters, such as the effect of shadows on neighboring buildings.

GIS is helpful for spatial analysis and offers various functions and physical spatial representations of an environment surrounding building digital models. BIM and GIS provide a useful platform for sharing location-based information and digital representations of architectural objects and environmental entities among stakeholders (Shirowzhan, Tan et al. 2020). BIM focuses on modelling buildings to their finest details, which consists of relevant microlevel data. At the same time, GIS focuses on generating macro-level information about the surrounding environment, such as the terrain and topographical environment, streets, and vegetation. As BIM and GIS features complement each other, a much more comprehensive system can be developed to integrate them. The integrated system can improve data-driven decisions during spatial planning and analysis, achieving smarter cities.

Rapid population growth has increased the demand to construct more buildings, predict the amount of land required and develop urban areas (Al Rifat & Liu 2022). As the buildings cast shadows, there is a need to investigate to what extent the shadow of a building may affect the blockage of sunlight to the surroundings. This leads to the development of different rules regarding the planning of the built environment by diverse planning bodies in an attempt to make living conditions comfortable for the inhabitants (Todorović, Živković et al. 2018). The BIM-GIS integration can be useful for sustainability analysis from building to city scale and particularly shadow analysis that influences the level of energy used by neighbouring buildings. Shadow analysis is useful to architects and urban planners since it helps understand the blockage of sunlight to the surroundings and the specification of solar panels' location for maximum energy gain. The integrated BIM-GIS model should be readily useable by the different stakeholders involved in planning facilities.

In addition to shadow and sustainability analysis from building to city scales, there are other city planning considerations that can be evaluated and simulated in digital twins of cities. For example, development assessments and planning approvals for the development and maintenance of buildings may take much time and require thousands of dollars due to less accurate information on building models and their neighbouring environment, including buildings and vegetation. Accurate digital twins of cities created from integrating BIM and GIS systems can provide the required information (Masoumi, Shirowzhan et al. 2023), for all these purposes to be used as planning support systems, saving thousands of dollars.

Some publications focus on the integration of BIM and GIS, with various workflows referring to interoperability issues and applications in different contexts. Each of the integration approaches has advantages and disadvantages (Fosu, Suprabhas et al. 2015). While different workflows are used to integrate BIM and GIS, there is a lack of evaluation of various workflows for making digital twins of cities for shadow analysis purposes in a 3D web environment. This chapter aims to focus on the web-based integration of a BIM model in a GIS environment to provide easy sharing of the integrated platform and a thorough comparison of several workflows for integrating BIM and GIS to make digital twins of cities, focusing especially on shadow analysis. Indeed, the research questions to be addressed are: (1) what would be the added value of integrating BIM data in web-based GIS for digital twins; and (2) what are the challenges of such integration, and is there any loss of information during each workflow of the integration?

7.2 Web-based BIM-GIS integration and digital twin concepts

There are different methodologies for integrating BIM and GIS developed by researchers and currently in practice in the industry. There are different platforms available for the integration of BIM and GIS, either open-source or licensed platforms. Current practice suggests there is a demand for BIM-GIS integration to adequately provide details of buildings and projects with respect to the geospatial context (Wang, Deng et al. 2019b). A digital representation of the physical assets is required to get insight into the city's infrastructure and landscape. The digital paradigm has the potential to rescue cities from expensive failures in the long term (TechwireAsia 2020).

Borrmann, König et al. (2018) discuss the advantages of BIM as a platform for the ongoing use of digital building models across the full lifespan of a constructed facility. BIM significantly improves information flow compared to conventional paper-based workflows (Borrmann, König et al. 2018), which are more prone to manual errors. A BIM model is created in a data-rich platform and is a parametric digital representation of physical objects. Various users can retrieve and analyse the views and data according to their needs. This information can be used to improve decisions in the delivery of the facility (Al Rifat & Liu 2022). Furthermore, enhanced visualisation of complex geometries is provided by BIM implementing parametric modelling as well as efficient data storage and information sharing and exchange (Haiying 2022). This approach considerably improves the coordination and quality of the design and engineering processes (Alexiadi & Potsioy 2012).

For urban planning, the detailed BIM models of public infrastructure can be linked with the existing plan and can be used to identify any potential problem associated with construction sequences. These sequences are required for supply chain management or logistics planning (Kaewunruen & Lian 2019) and monitoring the condition of the building (Matos, Rodrigues et al. 2022). BIM contributes to an extended asset life and enables a better estimation of future construction industry needs (Makkonen 2016). Building performance analysis and sustainability analysis may be done throughout the design phase since interdisciplinary information can be placed inside a BIM model (Chen, Cai et al. 2022).

As technology advances, a wide range of BIM tools suitable for planning, designing, and constructing built facilities are available in the market commercially or as open-source (Makkonen 2016). The BIM platforms support diverse file formats such as Standard for the Exchange of Product Data (STEP), Industry Foundation Classes (IFC), Standard Tessellation Language (STL), Initial Graphics Exchange Specification (IGES), Wavefront OBJ as well as design and drawing formats, such as DNG, DWF, DWG, and DXF. The leading BIM technology providers are Autodesk (2019), Tekla (2019), and Bentley (2019).

Even though BIM has several features that provide a number of advantages for the construction process, BIM adoption has been relatively slow in the industry. Reasons for slow utilisation include technical complexities for computability, data interoperability, and data integration among the model components. Well-defined construction process models and developing practical strategies can be used to overcome the barriers to implementing BIM (Alexiadi & Potsioy 2012), but using different modelling tools within a team and the necessity to migrate from one environment to another can be a big problem in BIM development. By using standards for data exchange, the data interoperability issue can be reduced (Eastman, Teicholz et al. 2011).

One of the considerable drawbacks of the current status of BIM is that even though it provides a detailed and enriching model of a construction facility, it lacks sufficient information regarding the surrounding context. Each process in the smart city life cycle can benefit if the digital model is connected with geographic features. For this study, a BIM model in Autodesk Revit was created, but even though Autodesk Revit can accommodate shadow studies, as a standalone model, it lacks the ability to represent the effect of shadow in the surrounding environment (Vandezande, Read et al. 2011). For an enhanced model, combining the BIM model with surrounding information is necessary. GIS is a powerful tool used for analysing geographic features (ESRI 2012), and the geographic features surrounding a construction project (e.g., a building) can be provided within a GIS environment (Ma and Ren 2017). The following section reviews the features of GIS.

7.2.1 Geographic Information System/science (GIS)

According to ESRI (2012), GIS connects geographic information and geospatial data with different attributes and descriptive information. Since all this information is viewed on a map, it provides an advantage over using spreadsheets or databases. Spatial analysis and maps can highlight patterns, point out the problem, and show

Table 7.1 Different open-source GIS tools

Software/release year	*Useful for application*	*Development platform*	*Software license*
Geoserver (Maurya, Ohri et al. 2015)	Map viewer, feature viewer or editor, 2D view	JAVA	GPL
QGIS (Lacaze, Dudek et al. 2018)	View, edit, analysis, and 3D view of data is possible with an add-in	C++, Qt4, Python	GPL
gvSIG (Terres de Lima, Fernández-Fernández et al. 2021)	View, edit, analysis (Mobile Applications), 3D view possible with a plugin	JAVA	GPL
ArcGIS (Soward and Li 2021)	View, edit, analysis, 3D view	C++, Java, Python	Proprietary
AutoCAD map	View, edit, analysis, 3D view, CAD, and GIS integration	C++	Proprietary
Bentley map	View, edit, analysis, 3D view, CAD, BIM, and GIS integration	C/C++, C#, .NET	Proprietary

connections between data that may not be significant in tables or texts (Bansal 2018). GIS and geospatial data are essential constituents of building smart cities in the digital age. GIS is integrated into the workflow for land management, urban planning, and transportation planning (Tao 2013).

The common GIS software platforms are proprietary software and open source, and this classification is based on the license type. Open-source software has free licenses that explicitly define users' legal rights with the freedom to operate, revise, change, redistribute, and access the source code (Tsou and Smith 2011). The proprietary software is brought from a vendor for a license fee. The open-source software for GIS concerning server applications is MapServer, and Geoserver. Desktop GIS such as Quantum GIS (QGIS) and gvSIG are also experiencing increasing numbers of users (Steiniger, Weibel et al. 2010). Table 7.1 provides more details about the different open-source and proprietary GIS systems that are used.

Badea and Badea (2018) compared open-source and proprietary GIS software packages. Since they can be modified, the promoted open-source software has gained momentum within the GIS community, but proprietary software is not free, and it is still the main form of GIS software used, primarily because it is easy to learn and use compared to open-source GIS software. For sustainability analysis, especially for the study of building shadowing problems on the surrounding environment, GIS as a standalone feature has its limitations (Wang, Pan et al. 2019a). However, integrating BIM and GIS is beneficial for shadow analysis. The following section introduces the concept of BIM-GIS integration.

7.2.2 BIM-GIS integration

Spatial data from the environment outside a building or the relevant city infrastructure is necessary for most construction or city development projects. While GIS can provide spatial information, it lacks the details and the material information about the building. Integrating GIS and BIM helps city and facility managers be updated about any changes and make informed decisions about various phases of the building life cycle and the smart city infrastructure. Effective management of the information using GIS and BIM can benefit design development and related decisions. Site selection, energy system design, structural design, traffic design, climate assessment, performance evaluation, and design authorisation are some tasks that can benefit from the integration (Isikdag, Underwood et al. 2008; Sergi and Li 2014; Kari, Lellei et al. 2016; Park and Kim 2016). In addition, the operation and maintenance phases have the most integrated application of BIM and GIS, where the integration can be used for hazard response, disaster management, safety and risk management, reaching sustainable development goals (SDGs) (Hawken, Rahmat et al. 2021), energy management, indoor or outdoor navigation, and facility management (Bianco, Del Giudice et al. 2013; Kang & Hong 2015).

Integrating BIM and GIS offers solutions to many current challenges related to construction logistics and improves the construction process. However, the integration process can be challenging and cumbersome and requires tremendous data storage, reflecting the inclusion of many buildings at the city level (Masoumi, Shirowzhan et al. 2023). This can be attributed to the dissimilarity between GIS and BIM in terms of their level of detail, spatial scale, geometry representation methods, and semantic mismatches (Karimi, Dickson et al. 2010; Isikdag, Zlatanova et al. 2013; Amirebrahimi, Rajabifard et al. 2016). Various efforts have been made to classify the BIM and GIS integration in previous works (Karimi, Dickson et al. 2010; Fosu, Suprabhas et al. 2015; Amirebrahimi, Rajabifard et al. 2016; Basir, Majid et al. 2018), which can be classified into three major groups: (1) application level methods; (2) process level, and (3) data levels. The following section summarises the different integration methods.

7.3 Research method

Reconfiguring or rebuilding (Karimi, Dickson et al. 2010; Knoth, Mittlböck et al. 2019), where software patches modify existing GIS or BIM, are the most common integration strategies at the application level. This strategy necessitates a larger monetary and time investment, and in terms of operating capabilities, the outcome is rigid (Basir, Majid et al. 2018). Irizarry, Karan et al. (2013) developed a BIM-GIS integrated model to improve visual monitoring of the building supply chain. A BIM-GIS optical module was used to depict the availability of materials in the study. This program extracts information from several databases throughout the internet to generate a real-time quote on doors and windows. GIS-based spatial analytics, such as network analysis and attribute analyses, were employed to provide an appropriate solution for managing supply chain logistics costs.

The web-based integration, like OWS-4 project by OGC (Lapierre and Cote 2007), employs Service Oriented Architecture (SOA) for the process-level technique, allowing integrated systems to collaborate on activities that need both capabilities while remaining active and unique. The following section provides an overview of the methods that have been proposed utilising the process-level technique.

Kang and Hong (2015) presented a software architecture for integrating BIM into a GIS-based facilities management system. The proposed program separates geometrical information from pertinent or associated attributes. The elements necessary for each use-case viewpoint are extracted and translated from BIM to GIS using the Extract Transform and Load (ETL) paradigm. The ETL idea collects homogeneous data from the source system, which is then converted into the right format or structure and placed in the warehouse. Rafiee, Dias et al. (2014) provided a step-by-step strategy for integrating BIM into a spatial information model using ETL.

Data sharing and information exchange between BIM and GIS is still a significant problem for BIM and GIS integration. Synthetic or semantic techniques are insufficient for transmitting semantic and geometric information from BIM to GIS and vice versa. Hor, Jadidi et al. (2016) devised a method based on semantic web technologies and the Resource Description Framework (RDF) to connect BIM and GIS and created the Integrated Geospatial Information Model (IGIM). This model can retrieve the processed dataset from GIS and BIM without using the RDS graph to create a direct relationship between terminologies. GIS-RDF and IFC-RDF graphs were generated using CityGML and IFC. At the semantic level, both graphs have been integrated, and the technology allows data from architectural elements to be linked to 3D/2D geographic data in the IGIM unified domain ontology.

Pauwels, Zhang et al. (2017) discussed semantic web technologies in the architecture business and construction industry. Semantic web technologies play a crucial role in information management where they come from hybrid and a range of systems, including GIS, BIM, energy management tools, and logic-based applications. Interoperability, linking across domains, logical inference, and proofs are the key application areas of semantic web techniques. The author's research stated that there is no solid proposal so far or any recommendation that solves the interoperability issues better than existing approaches. While they consider the linked data approach as the key method used by semantic web development, some scholars and practitioners suggest linked data is a fast-track approach in process-level integration. In considering logical inference and proof, it uses the logical basis of Web Ontology Language (OWL) and Semantic Web Rule Language (SWRL). Because this technology is at the top of the semantic web technology hierarchy, creating and deploying in real-world scenarios requires a lot more time and work. This strategy is more adaptable than the previous one. However, to ensure interoperability across these systems, the issues of integration must be handled at the fundamental data level in using this technique.

Several approaches have been developed for integrating BIM and GIS at the data level. For example, utilising an Application Programming Interface (API) on either side, connection technologies like ESRI ArcSDE provide data transmission between BIM and GIS applications. FME convert (van Rees 2014) is a translation/conversion technique that directly translates GIS and BIM files. Data is often translated between IFC and CityGML. FME is used to streamline the interpretation of spatial data between geometric and digital formats. The data transformation includes creating a single IFC file mesh, setting scale and coordinates onto the unique mesh, and writing the data to the CityGML format (Amirebrahimi, Rajabifard et al. 2016).

Isikdag and Zlatanova (2009) stipulated that manipulating data needs transforming both the geometric and semantic datasets. It is possible that the two environments are not aligned, preventing one of the datasets from being changed. The manual conversion or translation of IFC to CityGML usually entails the following steps: semantic filtering, external shell computation, building installation inclusion, geometric refinements, and semantic refinements. This is one of the most well-known conversion/translation frameworks for IFC and CityGML.

One of the constraints of the conversion between IFC and CityGML, according to Isikdag, Zlatanova et al. (2013), is missing information and lack of semantics. The characteristics' original meaning is lost even if the semantic information is complete after translation (Kang and Hong 2015) because natural geometric translation is not always possible. Converting from CityGML to IFC is more complex in terms of semantic contents and building geometry. Shared database access, direct data input, integrated data management, file translation, and formal semantics are the four data integration methodologies described by Mignard and Nicolle (2014).

Geiger, Benner et al. (2015) created a method to simplify the IFC model's complexity, considering semantics and geometry, for the IFC model to be compatible with one system. The study was done on IFCExplorer, a software tool for integrating, visualising, and analysing geographically referenced data. The key issue of this type of conversion is the missing semantics and difficulties of geometric conversion, as well as the translation of significant components and missed utilities and connections.

Even though they have benefits, some problems occur during implementation, such as data management, sharing, and topology functions (Amirebrahimi, Rajabifard et al. 2016). Basir, Majid et al. (2018) suggested that the future integration approach should have a unified model to address the integration complexities. A web-based integration can be a beneficial solution, allowing information to be shared quickly with all the stakeholders involved. This recommendation was supported by Akob, Hipni et al. (2019), who integrated BIM and web-based GIS in the infrastructure management of highway projects in Malaysia, which significantly improved overall project management there. This chapter proposes the data-level translation of BIM data to be suitable for seamless integration in the chosen GIS platform. The data-level integration methods, such as using different APIs, are easy and inexpensive (Basir, Majid et al. 2018), and in recent years, web-based GIS has been gaining popularity with the analytic properties of web-based GIS increasing (King 2019). It would be beneficial if the integrated platform used web-based GIS, as web GIS-based applications are easy to share with the stakeholders involved

(Akob, Hipni et al. 2019). The following section discusses the concept of a web-based GIS application.

7.3.1 *Technical attributes of 3D WebGIS and WebGL framework*

Web-based GIS can be 2D or 3D, but 3D enables better visualisation of all the features and the efficient and seamless sharing of ideas and concepts with the stakeholders and team members. The analytical side of 3D GIS is also beneficial as the 3D models can anticipate the actual environmental processes and are scalable, allowing models to study the impact from a range of areas, from single building studies to projects on a large scale (Schueren 2017). The client-server architecture enables the web distribution of GIS, allowing more convenient access to the database. HyperText Transfer Protocol (HTTP) facilitates client-server conversations. The client makes an HTTP request to the server, and the server responds with the desired data in an HTTP response (Mozilla 2019).

Standardisation as a guide is a vital aspect that supports the 3D web-based GIS. The Open Geospatial Consortium (OGC) defines a set of data models and web service standards that the GIS community uses to promote geographic data transmission and interoperability (Sayar, Pierce et al. 2005). The most often used OGC web frames for web mapping are Web Map Service (WMS), Web Feature Service (WFS), and Web Coverage Service (WCS).

WMS creates a digital picture file with georeferenced data maps that may be shown on a digital screen. The maps are commonly represented in a visual format like JPEG or PNG or as vector-based graphical components in SVG format. The client features of WMS can be supplied directly via the Uniform Resource Locator (de La Beaujardiere 2006). At the functionality and feature property levels, WFS gives direct access to geoinformation, while GML is a format for storing and serving spatial data. Users may now query features using both spatial and non-spatial limitations. Transactions for establishing, updating, or removing features and a method for locking features during a transaction are among WMS's additional capabilities (Sayar, Pierce et al. 2005; Vretanos 2005). WCS provides access to multi-dimensional coverage data to promote networked geographic data interchange. After providing the access data with their thorough prescription, the service delivers data with its original semantics that can be interpreted and analysed (Whiteside and Evans 2007). OGC standards where WMS is used for providing georeferenced map images and WFS for querying the geographical features are suitable for developing 3D Web GIS applications (Manolova 2018).

The 3D web-based GIS uses the WebGL frameworks (Mete, Guler et al. 2018), and high-level languages like C or C++ are used to develop the 3D graphics applications. They are compiled into binary data for a specific platform. This could be the reason for the application complications appearing on different operating systems and graphics sharing due to the required plugins (Matsuda and Lea 2013). Modern browsers have improved their rendering features by creating interactive components through the canvas presented in HyperText Markup Language (HTML) and the interactive functionalities of JavaScript (Dirksen 2013).

The benefit is that it allows users to use the applications without installing any web browser plugins.

The WebGL framework is developed based on the Open Graphics Library (OpenGL) and the open standard widely used in computer graphics and video games, enabling faster rendering (Matsuda and Lea 2013). WebGL compilation is done by combining JavaScript and OpenGL Language Embedded Systems (GLSL ES), a shading program consisting of a vertex and a fragment shader (Eck 2016), reflecting the main difference between traditional dynamic HTML pages and HTML pages with WebGL content.

Examples of the different WebGL-based frameworks are Cesium, OSM buildings, three.js, iTowns, and ArcGIS online, which are used to visualise 3D models and animations that can be used for the development of 3D web-based GIS applications (Manolova 2018). Cesium and ArcGIS online are chosen for the analysis in this chapter since they are the most advanced WebGL-based frameworks that can be used in BIM and GIS integration (Deng, Gan et al. 2019). Another benefit of Cesium ArcGIS online is that it provides an API for other software/web-based services, which can be used to create more advanced and powerful web-based GIS applications. The next section gives an outline of Cesium and the various methods to integrate the 3D BIM model in Cesium.

Cesium is used to transform 3D geospatial data into 3D content (Cesium 2019a). The main feature of Cesium is that it has a higher-quality of 3D tiles for the representation of BIM, CAD, and Design models (Cesium 2019b). Cesium created 3D tiles, an open specification for streaming enormous heterogeneous 3D geographic resource sets. Buildings, trees, point clouds, and vector data are all examples of 3D material that may be streamed using 3D tiles. When the 3D model is created successfully, the 3D BIM model is imported to the Cesium ion account. The Cesium ion platform optimises the uploaded contents, tiles them into a web-friendly format host in the cloud, and streams them to any device. The BIM-supported formats used by Cesium ion for 3D tiling are glTF (.gltf,.glb), Filmbox (FBX), and KML/COLLADA. Table 7.2 shows different data formats and their features.

7.3.2 Web-based GIS applications

Improvement in technology catalysed the transition of the traditional desktop to web-based GIS, which improves the availability and dissemination of geoinformation to all the stakeholders involved; this transition results from web technology development (Manolova 2018). Similarly, like traditional GIS platforms, web-based GIS has the capability and computational ability to process geographical information. The only difference between traditional and web GIS is that the traditional one is based on the desktop, while the latter is based on the web (King 2019). The evolution of web GIS can be seen from presenting basic maps provided in raster formats to dynamic maps generated directly from spatial databases. Due to the above development, web applications enable interactive data representation, allowing users to view, query, analyse, and freely download georeferenced data (de Paiva and de Souza Baptista 2009).

Table 7.2 Comparison of different data formats and their features

Data format	*Features*	*Information stored*
GL Transmission format (*glTF*) (Chen, Shooraj et al. 2018, Blut, Blut et al. 2019)	A web-based JavaScript framework is supported by this format, which is useful for small file sizes, rapid loading, and run-time independence. glTF assets are JavaScript Object Notation (JSON) files plus supporting external data.	JSON formatted file contains node hierarchy, materials, descriptor information, and animations (Izawa and Koga 2019). Binary files (.bin) are used to store geometry and animation data. Image files (.jpg,.png) for textures.
Filmbox (FBX) (Chen, Luo et al. 2021)	FBX is a platform-independent 3D authoring and interchange format. Proprietary format, difficult to load directly in a web browser.	The 3D scene is stored in binary or ASCII format. The data for the 3D scenes are about the camera, lights, meshes, animation, and textures.
Keyhole Markup Language (KML) (ESRI 2019)	Open Geospatial Consortium (OGC) standard is an XML-based file for the storage of geospatial data and related material. A useful format for non-GIS applications, including Google Earth and ArcGIS Explorer, it is composed of feature and raster elements and is used to create maps in the GIS system.	XML-based language to visualise geographic data in 2D or 3D. KML comprises points, lines, raster imagery, graphics, and attributes.

Cesium has a javascript library used in applications for creating 3D maps. The platform supports various spatial data sources and implements the primary OGC web services like WMS and WFS, which makes it a viable option for building professional geospatial applications and creating 3D city models (Schilling, Bolling et al. 2016). The significant advantage of Cesium is that the most common BIM file formats can be loaded in Cesium, but to be rendered as a 3D format, they need to be converted into GL transmission format (glTF). Besides the well-performing and precise maps, the support provided by the Cesium development team and the contribution from the user community ease the process of developing applications in Cesium (2019b). The advantage of Cesium is its support for OpenStreetMap, Bing Maps, ArcGIS Mapserver, as well as 3D BIM formats, such as glTF, Wavefront, COLLADA, GeoJSON, KML, and LASer (Manolova 2018).

ArcGIS Online is useful for collaborative tasks to create, update, and share files and services. ArcGIS is a proprietary platform developed by ESRI. One of the advantages of ArcGIS online is that the various scenes created by the desktop software (City Engine, ArcGIS Pro) can be uploaded to visualise and analyse on the web. The graphic interface of the portal is intuitive, and 3D visualisation is supported (Cecchini, Magrini et al. 2019).

7.3.3 *Earlier attempts for BIM and GIS integration using web-based GIS application*

The integration of 3D BIM models in Web GIS has been extensively researched in diverse fields (Manolova 2018). The following section provides an overview of several approaches to web visualisation. For example, Schilling, Bolling et al. (2016) developed a method for the real-time streaming of extensive and highly detailed CityGML data based on the open standard. Researchers used GL transmission (glTF) format to encode the 3D models and 3D tiles to associate the glTF objects with their attributes and stream the 3D data. For transferring multiple 3D models in a single request and rendering them efficiently using batches, the batched 3D model (B3DM) is implemented within the 3D tiles concept (Graphics 2019). To improve data exchange and interoperability, various 3D models of buildings and urban objects are visualised in Cesium (Schilling, Bolling et al. 2016).

Prandi, Devigili et al. (2015) did similar research by developing an approach for accessing extensive CityGML data. The method is based on web services by separating the semantic and geometric parts of the 3D models, and the researchers used NASA World Wind (NWW) and Cesium to display the geometries. The highlight of both the web technologies is it's open-source virtual globes that allow the efficient representation of 3D city models. Java applet, with an optimised custom-rendering engine for 3D buildings, is utilised by NWW. The CityGML data are loaded in a custom servelet and transformed into the C3D file format readable in NWW. A promising solution is to integrate the CityGML models in Cesium using C3D and glTF. By using WFS service, the semantic information about the 3D models can be retrieved from 3DCityDB.

Chaturvedi (2014) developed a web application to visualise virtual 3D city models and perform 3D buffer analysis using WebGL. Decision-makers and field workers can use the 3D buffer analysis tool for evacuation planning. The workflow for the representation of CityGML started with extracting data to the Keyhole Markup Language (KML) format, which is readable by many JavaScript libraries. To generate the 3D buffer zones, several algorithms were computed to create 3D buffer zones around the affected buildings on the fly and determine the urban projects that are affected by the buffer zones.

For public space management, Manolova (2018) created a framework for integrating a 3D BIM model of a bridge in Web GIS. The model was created on Autodesk Inventor. The BIM model is converted into an OBJ file, which can be integrated into a 3D web application created using the three.js framework. Liu, Zhang et al. (2019) created a Cesium-based framework to visualise BIM data on the web. The authors developed the framework to make a sizeable BIM and data lightweight for visualisation in the research. Trimming BIM data is done through the model's hierarchical partitioning, and rendering the BIM model is done using three.js. The integrated platform has many add-in functions such as plane measuring, material editing, and semantic attributes' searching.

Cecchini, Magrini et al. (2019) presented the workflow to implement a web-based application where the BIM model can be represented in a 3D GIS.

The platform should be able to store, handle, and display the information related to the energy consumption of the building. The web application is developed using ArcGIS online as the base. Geoff (2017) studied mago3D, a 3D web-based GIS platform that Gaia3D developed. The platform's advantage is that it is web-based, so installing any other software to run and handle massive, complex BIM objects is unnecessary. It has an engine that is based on an internal format. To visualise the BIM element in a geospatial context, it integrates open-source virtual globe Cesium and NWW.

Chen, Shooraj et al. (2018) examined the BIM example provided by Bentley Systems, a company in Pennsylvania, United States. Selection and queries can be made for each BIM component in the example. However, the integration workflow is not open. Moreover, (Chen, Shooraj et al. 2018) proposed a workflow to convert the BIM data into 3D tiles for visualisation in Cesium using open-source libraries and converters available on the internet. However, the proposed workflow is resource-intensive and time-consuming for the visualisation purpose of 3D BIM in Cesium.

Ignatius, Wong et al. (2019) stipulated a workflow used for Singapore's digital twin development. The development of the digital twin is done primarily by mapping and converting IFC-BIM models into CityGML. The algorithm is developed for mapping virtual Singapore with an interactive dashboard, but the process is time-consuming (National Research Foundation 2020). Errey, Noonan et al. (2019) developed a digital twin for the NSW government with the help of the open-source platform Cesium and the in-house developed platform Teria and Magda. The platform integrates a live API feed for public transport energy production and air quality (Services 2020).

There are studies on BIM-GIS integration for various purposes in the construction domain. For example, after reviewing current BIM-GIS integration approaches, Zhao, Mbachu et al. (2022) proposed a web-based integration method for a mega construction project. In another study, Bansal (2021) proposed a novel framework for using BIM-GIS integration for applications covering the entire life cycle of a building. While these studies are all valuable in various domains and for different reasons, there is a gap in exploring shadow analysis in BIM-GIS integration workflow for applicability to digital twins of cities and impact assessment of new urban developments.

7.3.4 Need for comparison of workflows

There are different applications for integrating a BIM in web-based GIS in the construction lifecycle, and there are different integration methods available in the literature, but there have been few attempts to compare the integration workflow between Cesium and ArcGIS online. However, El Haje, Jessel et al. (2016) created a web-based solution for visualising 3D cities using Cesium and ArcGIS online, and a comparison of the two results was made for visual quality. El Haje, Jessel et al. (2016) concluded that the ArcGIS-based solution has more built-in capabilities, but CesiumJS is a compelling platform for 3D visualisation. Cesium JS-based application performance varies from browser to browser. Stähl (2017) conducted a comparative study between Cesium and ArcGIS products where a web-based application is developed using Cesium and ArcGIS online API to show the settlement development of the city of Zurich. Both studies compare the integration of the 3D city database into Web GIS, but they do not consider the integration of the 3D BIM model developed in Autodesk Revit.

7.3.5 *Need for shadow analysis*

Rafiee, Dias et al. (2014) declare that shadow and view studies may be performed utilising geometric and semantic data from georeferenced 3D BIM and other geospatial components. This analysis provides better design choices in urban design, but the method requires higher computing power and a more complicated process. Consequently, the authors suggested that providing a web-based solution would simplify the process of sharing the integrated result. Further adding to the analysis of shadow is an inbuilt application of CesiumJS (Cesium 2019b). Similarly, the effect of the shadow is a default feature for the 3D scene that is shared using ArcGIS online (ESRI 2019); it would be beneficial if the shadow analysis could be compared using the two platforms for stakeholders involved to select a suitable workflow according to their needs.

The benefits and challenges of integrating BIM into GIS are discussed, referring to why the integration of BIM in WebGIS is relevant for the sustainability analysis, and what the gaps are in the current literature. To address the integration, the Web GIS considered is Cesium, an open-source virtual globe developed in JavaScript, and ArcGIS Online, a proprietary Web GIS developed by ESRI. The aspect used in the study is a comparison between BIM-GIS integration using Cesium and ArcGIS online. Once the integration is done, the next step is to analyse the effect of the shadow on the surrounding environment. To understand the features of integrating BIM and GIS in web-based GIS, a 3D BIM model is georeferenced in web-based GIS. The proposed method is organised as follows: design of the residential project; integration of BIM and GIS; application of Cesium and ArcGIS Online; shadow analysis; and a comparison of different workflows, which will be discussed in the following sections.

7.3.6 *Design of residential project*

Since the research is about comparing different workflows used to integrate the BIM model into GIS for sustainability analysis, the web GIS platform selected for the integration is Cesium and ArcGIS online, which are pursuing varying approaches to processing, data modelling, visualisation, and symbolisation. A fair comparison can only be made by implementing identical use cases. Therefore, it is best to make a comparison by implementing a standard 3D BIM model in both Cesium and ArcGIS online. The 3D BIM model is developed in Autodesk Revit. The 3D BIM model was developed for a proposed house located at 50 Craik Avenue in the suburbs of Australia.

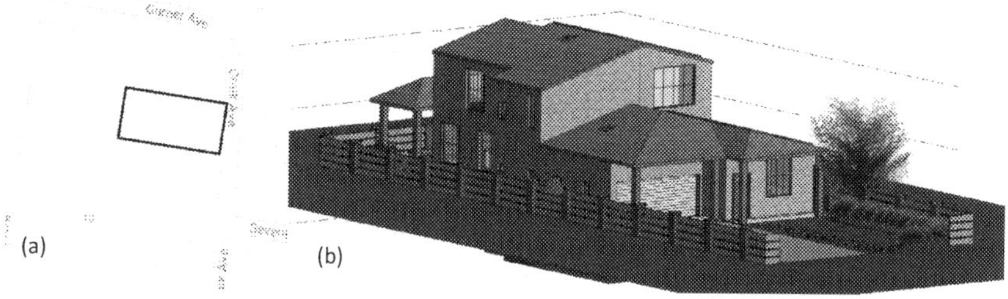

Figure 7.1 (a) 50 Craik Avenue sourced from Google Maps; (b) 3D BIM model developed in Revit.

The benefit of using Autodesk Revit, compared to other 3D BIM models, is that Autodesk can provide support for the entire lifecycle study of a BIM model (from 3D BIM to 7D BIM). Developing a 3D model in Revit has the additional benefit that it can be used in different Autodesk software suites used for other phases of BIM. The strength of Revit lies in the coordination of documents, their instant update, scheduling, and the ability to view when any changes are made to the model (Khemlani 2004). The Revit file is created as a construction project in Revit, and it has an extension of *.RVT. Autodesk Revit is able to export the *.RVT file into a different file format that includes *.FBX, IFC, and DWG.

The level of detail required for each design stage of BIM was initially not clearly defined by the construction engineers. The American Institute of Architects (AIA) and BIMForum introduced the level of development concept (LOD) to add clarity to the building process. The LOD concept specifies the geometries and attached information that should be represented at the various levels (Yoders 2013). BIM models can be classified into five levels. The BIM models become more detailed and informative at each level with the increasing LOD (BIMForum 2013). In this chapter, a LOD 300 3D BIM model was developed for the Craik Avenue house, designed according to local council rules. Figure 7.1 shows the location of the selected building, which is used as a case study, and the model is created in Revit.

7.4 Integration of BIM and GIS

The integration of 3D BIM model into a web GIS environment is done through several workflows. A theoretical background in 3D web-based GIS and the framework supported by 3D Web GIS can be useful for understanding the workflows. Figure 7.2 shows the timeline of scholars' attempts to integrate BIM and GIS and the recent trends. It shows that the integration of GIS-BIM has been focused on sustainability and energy management to address modern urban challenges in recent years. Wen, Ren et al. (2021) reviewed the relevant literature and suggested that the integration is useful for reducing the effect of construction activities and building developments on the surrounding environment from the sustainability perspective. Wang, Pan et al. (2019a) highlights that the recent reasons for the integration are connected with energy management and urban governance, including disaster

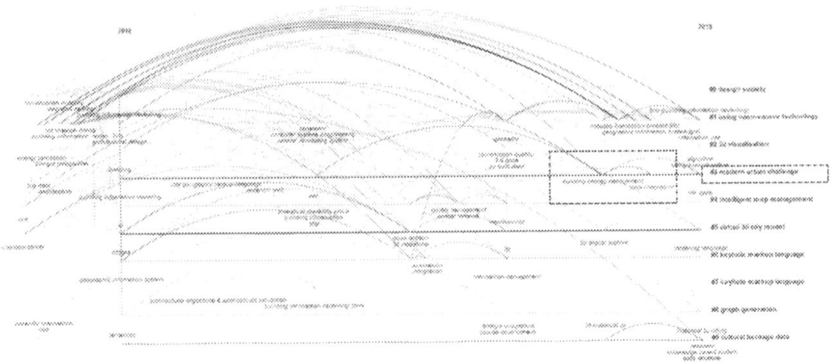

Figure 7.2 GIS-BIM integration topics over time in the literature.

management. Also, the main issues discussed in the literature are information management, BIM-GIS collaboration platforms, and modelling languages. The challenge of information management and the quality of the integration outcome require many practices and workflow development. The following sections present different workflows designed and tested using Cesium and ESRI GIS products.

7.4.1 Implementation of Workflow 1

The 3D BIM model is created in Revit as an *.rvt file. The 3D model can be exported into *.fbx using the inbuilt Revit to FBX export function of AutoCAD Revit, but the exported 3D file lost its texture and material properties. To overcome this issue, the Twinmotion Revit FBX export plugin was used. The Twinmotion export plugin is based on the unreal engine, and it reduces the size of the *.fbx file significantly (Twinmotion 2019). Figure 7.3 shows the *.fbx format that facilitates higher fidelity data exchange as one of the key components of a digital twin as well as the interoperability between 3Ds Max, Maya, MotionBuilder, and Mudbox. FBX can be represented on a disk as either binary or ASCII data (Autodesk 2019). In Cesium ion, the 3D tiling of FBX file is executed with the assistance of an open-source FBX2glTF converter developed by Pär Winzell and the team at Facebook (Cesium 2019b). This uses glTF as an intermediate format for 3D tiling. File size after exporting from Revit is 12.95 MB, 4.5 MB, and 1.526 MB with an *.RVT format, using an Autodesk FBX export plugin, and a Twinmotion FBX export plugin, respectively. Less data storage means greater ease of uploading and viewing 3D tiles in Cesium (Cesium 2019b).

The 3D model in FBX is hosted in the Cesium ion account, which serves as a hub to represent the 3D assets. Draco compression of the 3D content hosted is conducted, Draco being a glTF extension for mesh compression, which uses an open-source library developed by Google to compress and decompress 3D meshes to reduce the size of 3D content. Draco improves the efficiency and speed of transmitting 3D content over the web by compressing vertex positions, colours, texture

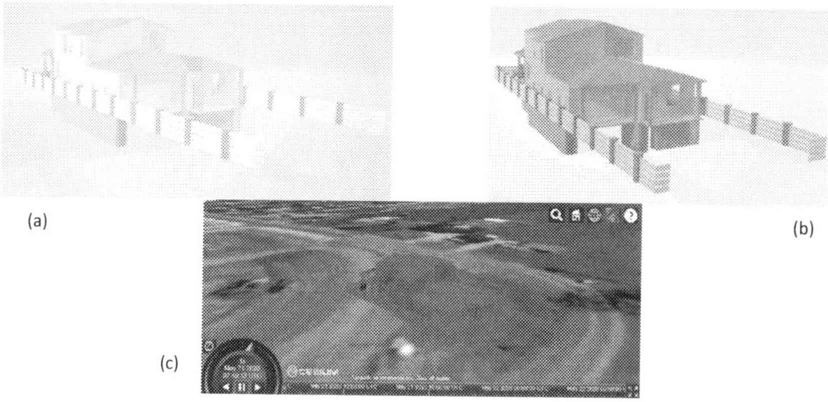

Figure 7.3 (a) The FBX file is converted using Autodesk Revit inbuilt FBX exporter visualised in Microsoft paint. (b) The FBX file is converted using the Twinmotion FBX export plugin and represented in Microsoft paint. (c) The figure shows the 3D model in FBX format georeferenced to the location.

coordinates, and any other generic vertex attributes. This decreases the size of files and assists in faster streaming of 3D content.

7.4.2 Implementation of Workflow 2

In this workflow, the Revit file is converted into FBX first, and then the FBX file is converted to a KML file. Similar to the conversion of *.RVT to *.FBX in Workflow 1, the conversion of the *.RVT file to *.FBX is done using a Twinmotion plugin. Then *.FBX format is converted to KML using FME workbench by Safe software. FME is a data transformation software that supports the formats that are used in BIM and GIS. The data translation software can read and write more than 300 data formats, including FBX and KML. By using FME, the translation of spatial datasets is relatively easy as it has a lot of prebuilt transformation tools, and there is no need to write codes to customise the workflow (see Figures 7.4a and 7.4b). The look of the BIM model will be simplified by utilising the FME workbench with the FBX file to eliminate the system's automatic overlapping shadow of pieces. The main goal of data simplification is to decrease data loading and provide a visual representation of a higher quality.

In this case, the ideal geometry to write out the KML file is as a single mesh. Inbuilt transformer functions facilitate these conversions. As seen in Figure 7.4c, the transformer functions used in the translation process are Triangulator, MeshMerger, KMLPropertySetter, and LocalCoordinate SystemSetter. Their functions are as follows: (1) Triangulator is used to break down the FBX geometry into a mesh; (2) MeshMerger, is used to unify the mesh generated by Triangulator into a single output mesh. This allows for the efficient storage of geometry and faster translation of data; (3) KMLPropertySetter and Local Coordinate System Setter are used to set the relevant properties for the KML file. Then, the output KML data is

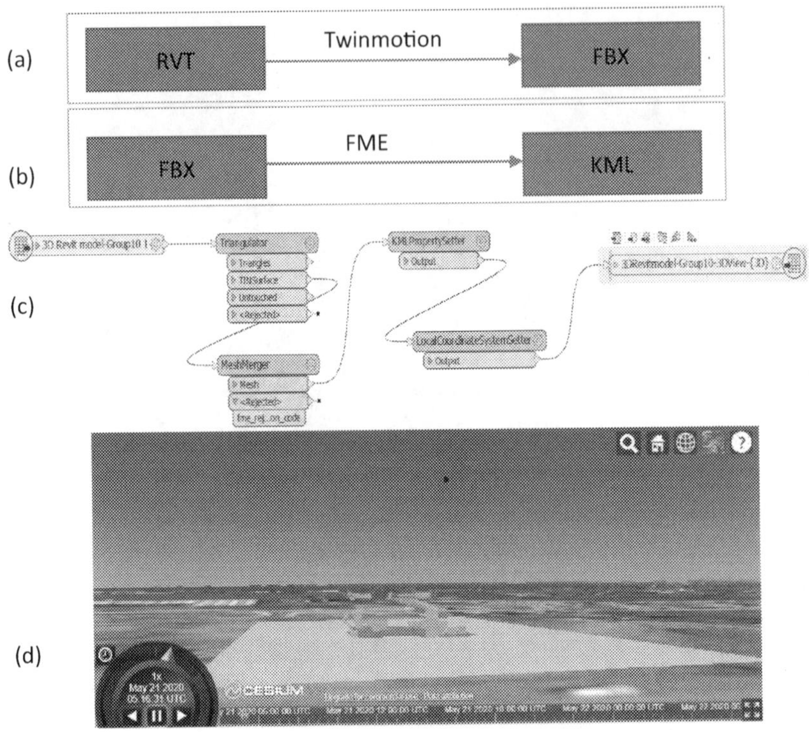

Figure 7.4 (a) Conversion of RVT to FBX in Workflow 1; (b) conversion of FBX to KML in Workflow 2; (c) the conversion of *.fbx file to *.kml using FME. (d) The figure shows the 3D model in KML format suffered data loss during the conversion process georeferenced to the proposed location.

uploaded into the Cesium ion account, and then the 3D model is displayed in the Cesium sandcastle. As seen in Figure 7.4d, the outcome of this BIM-GIS integrated workflow in Cesium suffers from considerable data loss.

7.4.3 *Implementation of Workflow 3*

In this workflow, we used the twinmotion export plugin to convert the RVT file to an FBX file format, which is the first step in this workflow. The FBX file is then imported into a blender, and the model is exported as a *.glTF file (see Figure 7.5). Blender is a useful 3D creation suite that supports the entire process of the 3D pipeline – modelling, rigging, and rendering. Blender has an inbuilt *.glTF export plugin. In *.glTF, 3D models may be sent and loaded in online and native apps. The *.glTF reduces the size of 3D models and the amount of processing time required to unpack and render them. Because glTF's internal structure mirrors the memory buffers frequently used in graphics cards, those materials may be supplied to desktop, web, or mobile clients in real-time and shown with little processing. Then, the converted model in *.glTF format is represented in Cesium ion. Since 3D tiles for

Figure 7.5 (a) The figure shows the 3D model in Blenders' render view; (b) the figure shows the 3D model in glTF format georeferenced to the proposed location URL: https://tinyurl.com/y8prajcr; (c) the figure represents the shadow features; (d) the figure represents the shadow cast by the house on the surrounding environment at a particular time and date highlighted in the figure.

Cesium are built on *.glTF, it makes a natural fit for the representation of 3D data in Cesium ion (see Figure 7.5).

7.4.4 *Shadow analysis using Cesium*

Out of the three data formats used for the study, *.glTF has a better quality of representation. For the study of shadow analysis, *.glTF is selected. The viewer is the basis of any Cesium application; It is an interactive globe with lots of functionality. First, a variable for the viewer is created and attached to the div with ID Cesium contained. The following line of code represents the viewer: var viewer = new Cesium.Viewer ("cesiumContainer").

The scene in Figure 7.5 has default widgets, and further, the viewer can be configured to include or exclude more features by passing in an options object as a parameter. By default, there is a Boolean operator "shadows" to determine the shadows that are cast by the sun and the effect of those shadows.

7.4.5 *Implementation of Workflow 4*

In this workflow, we used two ESRI products for the integration of the BIM model (Figure 7.6) into a GIS environment. The 3D environments are 3D web scenes from ArcGIS Pro and City Engine. To import the *.rvt file into ArcGIS Pro, we converted it into a multipatch feature using ArcGIS Pro (Figure 7.7). Multipatch features are GIS object representations where the boundary of the 3D object is shown as a single row in the database through a collection of patches. The features stored by patches are texture, colour, transparency, and geometric information. ArcGIS Pro is a proprietary GIS desktop application by the ESRI that supports the creation of 3D scenes. The initial step was to add a folder connection in ArcGIS Pro

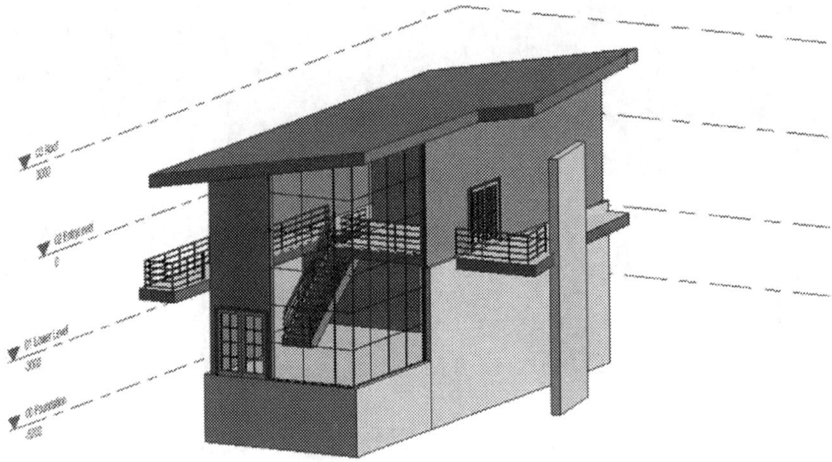

Figure 7.6 3D BIM model developed in Revit.

Figure 7.7 3D BIM model exported in ArcGIS pro.

with the Revit database. Once the connection is made, the Revit file can be added to a scene, and since the Revit project file does not have a geo-location specified during modelling, the BIM model will move into an arbitrary location in the scene. Then, we georeferenced the file manually with the inbuilt georeferencing tool. The model's further adjustment is done using a rotate, scale, and elevate to ground function to correctly position the scene's building.

In the second integration approach using CityEngine (Figure 7.8), the first step was to create map data as a base image. Then, the next step was to import the multipatch data created in the previous step from the geodatabase using the built-in import function. It was observed that the multipatch data came through as correct,

Figure 7.8 Building georeferenced in CityEngine.

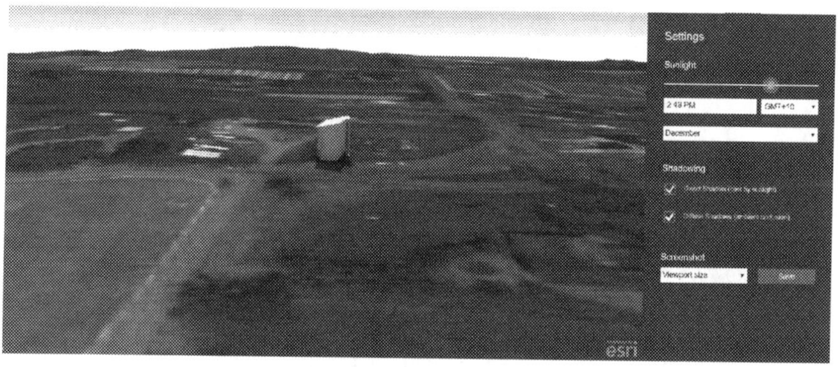

Figure 7.9 The scene represents the CityEngine web viewer with the shadows.

but the symbology did not come through. To customise the symbology, a default rule package available from ESRI is used. The rule package is a compressed package containing CGA rule, the file with several rules defining how a building geometry is created), which contains the referenced assets and data (ESRI 2019b).

7.4.6 Shadow analysis

The method discussed in the above section is for integrating BIM in Desktop GIS (Figure 7.9). The scene in CityEngine can be displayed as a web scene and can be exported as a 3ws file. A 3WS file is a custom web-optimised format developed by ESRI that can be shared on ArcGIS online and viewed in the CityEngine web viewer.

7.5 Results

In the case of Cesium, the integration of BIM and GIS is done using three different workflows. All the files are exported from a Revit file of size 12.6 MB. The different workflows are compared, as shown in Table 7.3.

Table 7.3 Comparison of the representation of BIM using a different workflow

BIM file format	*File format in Cesium*	*File size after conversion*	*Other tools required for data conversion*	*Quality of 3D visualisation*	*Analytics (measuring distance, area, and clipping)*
RVT	FBX	1.49	Twinmotion	Basement walls appear translucent	Currently a paid service in Cesium
RVT	KML	1.22	Twinmotion, FME	Only the furniture and fixtures are visible on the integrated product	Currently a paid service in Cesium
RVT	glTF	3.63	Twinmotion, Blender	Better visual representation compared to other workflows	Currently a paid service in Cesium

The analysis shows that the workflow using *.glTF offers a better visualisation quality, where the loss of semantics is less than with other solutions. Integrating the BIM model in GIS using ESRI products is time-consuming and complex as the translation is done using two pieces of software. It is observed that after the integration of 3D BIM in ArcGIS online, there is information loss as it is being observed that the balconies are missing. The results observed while doing the integration in both platforms are summarised in Table 7.4.

Cesium and ESRI products support the analysis of shadows. In Cesium, to implement the shadow analysis feature, special codes need to be written. In the case of ArcGIS online, shadow analysis is a default feature.

7.6 Discussion

A comparison between the output of different workflows is necessary to provide a holistic view of the different workflows used to integrate BIM in WebGIS. The bibliography analysis shows that the number of papers focusing on the integration of BIM and GIS is increasing, as shown in Figure 7.10. In addition, the figure shows that the GIS applications and literature are much wider than BIM, so this can help to appreciate the process of technology development, examine the applications offered by GIS scholars, and use them for BIM or DT development where appropriate.

7.6.1 Comparison of different workflows in Cesium

In the case of the FBX workflow, there is no need to use third-party software or API for further conversion of the 3D file format. The reason is that FBX is a common

Table 7.4 Comparison of Cesium and ESRI products for integration

Platform	*Ease of integration*	*View quality*	*Analytics*	*Sustainability tools for visualisation*	*Complexity of workflow*	*Material selection*
Cesium Needs a separation line between the two rows	As Cesium supports most of the 3D BIM formats, the process of integration is seamless.	As it is based on HTML 5, the view quality is high	Analytics such as measuring distance, and area, and clipping features are paid services	Dynamic shadow analysis can be done	Less complex	No material selection
ESRI Products	To represent the BIM data in ArcGIS Online, it had two translation steps involved where two different GISdesktop software (ArcGIS Pro and CityEngine) are involved	View quality is on par with Cesium	ArcGIS Pro has an inbuilt analytics feature for measuring distance area clipping. But once the scene is shared in ArcGIS online, it does not have the above-mentioned analytics features	Dynamic shadow analysis is possible with the added option to diffuse shadows	More complex and time-consuming	Material selection and display of the materials is possible

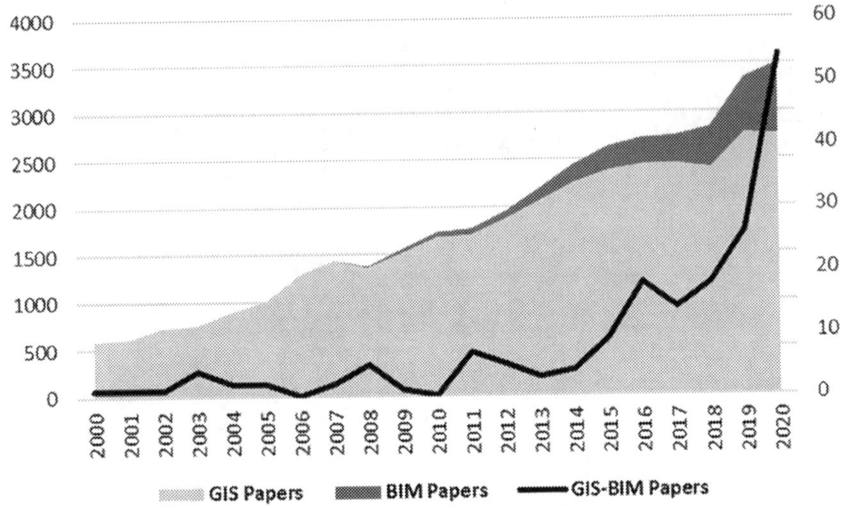

Figure 7.10 An illustration of the growing literature on the GIS-BIM integration theme over the Years

3D file format supported by most current 3D modelling software, and Autodesk Revit supports the export for the integration of the 3D model. One of the main disadvantages is that when the FBX file is loaded into Cesium ion, there is a semantic loss of some of the material. In this case, it was for the basement wall. It occurred because the Cesium ion interpreted the translucency property of the basement wall in FBX format as a true representation.

In the case of the workflow using the KML format, KML is supported for visualisation in web-based 2D and 3D earth browsers. It has an added benefit that the converted file in KML format can be represented in other web-based earth browsers. However, there is a semantic loss after integrating the 3D BIM model on google earth. As observed in the case of the FBX file format, the KML format integrated into Cesium experiences a semantic loss, especially for the basement wall.

In the case of the *.glTF format, it is observed that the representation of the 3D BIM model is more accurate without any semantic loss for the building element. Blender is used for the conversion of the 3D model into the *.glTF format. In a blender, customisation of the imported model is possible so the visual quality of the 3D BIM model can be improved. Since Cesium is based on the *.glTF format, the BIM model representation is better than the other two file formats. One of the disadvantages of the *.glTF model is that it has a larger file size compared to the other two formats. Currently, Cesium does not have any analytics features, including features to measure distance, area, and clipping. Special codes need to be compiled to include these features, which is time-consuming.

The advantage of Cesium is that it is a 3D web GIS suite based on a web GL framework used for 3D visualisation on the web. Cesium also supports integrating BIM formats such as FBX, KML, *.glTF for 3D tiling. In this research, the BIM model is created in Autodesk Revit (RVT format) and changed to different above-mentioned BIM formats supported by Cesium. As discussed in the section Comparison of the different workflows used in Cesium, the conversion of RVT files to FBX, KML, *.glTF involves multiple software packages. The most effortless workflow will be integrating FBX into Cesium as it involves only one software and translation step. Meanwhile, the other two workflows (Workflows 2 and 3) are complex, involving multiple software and translation steps that take more time. Since Cesium is based on the *.glTF format, the representation of the BIM model using the *.glTF format results in the integration of BIM and GIS without any semantic loss. Cesium does not have an inbuilt analytics option (Property to measure the area, distance, and clipping of scene). Cesium has a Boolean operator to visualise the shadow cast's dynamic effect by the building on the environment, which can be called by writing the necessary codes.

There might be various workflows to integrate the BIM data in GIS using Cesium to be efficient. However, using other commercial GIS tools may shorten the integration process because it involves two software suites (ArcGIS Pro and CityEngine). It is observed that when the BIM data is imported into ArcGIS Pro, the balcony in the imported model is missing. Another feature supported by ESRI is material selection. Individual elements in the BIM model can be selected, and the material properties are displayed. Shadow analysis is a default feature in ArcGIS online. To render the BIM model, it should be exported to the CityEngine rule package, which is time-consuming and complicated. The most significant advantage of using ESRI products is that they are developed to perform analyses related to GIS applications, so ESRI products have an inbuilt option for analytics, like measuring distance and area and querying data. ArcGIS online is user-friendly compared to other tools with interactive and self-explanatory widgets. Shadow analysis is an inbuilt option, and ArcGIS online also has the option to diffuse shadows.

7.7 Conclusions

The integration of BIM and GIS is a serious demand from the construction industry, and preparing integrated models for digital twins is a trending topic. While previous studies attempted to promote BIM and GIS integration, many integration challenges and different integration methods have not been covered entirely. The implementation of required tasks can be challenging and depends on the complexity of BIM and GIS cases. Due to the BIM's lack of spatial data analysis features, there is a need to deploy the building information or construction details in web-based GIS. In a nutshell, BIM and GIS are incompatible with regard to the modelling environment and the reference systems. BIM models have a local coordinate system, while GIS can be used to georeferenced the spatial data. BIM contains semantically rich and detailed information about building elements. At the same time, GIS focuses more on spatial analysis and geographical coordinates. A comprehensive system can be developed where detailed, semantically rich information is connected to its exact location. This chapter has examined the integration of BIM in an open-source GIS called Cesium in conjunction with ArcGIS online and shadow analysis.

Since there are a variety of BIM modelling platforms, the design techniques used in BIM differ significantly from platform to platform. The BIM designer is free to choose the platform for a 3D model of a building. As a result, it is difficult to find consistent 3D data and process it accordingly. The 3D data has to be processed beforehand for the integration of BIM 3D models in web-based GIS.

Shadow analysis plays an essential role in a different sector. One of the industries that can benefit from shadow analysis is the energy sector related to solar panels. Solar panels are essential renewable energy sources and have been gaining popularity recently. For solar panels mounted on the rooftop of a building, geometrical aspects of the building are necessary for optimal solar energy gain, and it partly depends upon the effect of shadow falling above the solar panels. Furthermore, city planners and architects can use shadow analysis to make a suitable decision in planning the geometry of the building.

A few limitations were experienced during the experiments. There was no inbuilt analytics feature in Cesium or some key tools at the time, which hindered achieving the full potential for BIM-GIS integration unless both Cesium and ArcGIS online could extend support to the IFC file format. The IFC file format is the universal standard data format of BIM. It also allows interoperability with different software that uses the BIM model in the construction industry. The workflow used for converting the BIM model to Cesium can be used to develop a 3D web application. The web application is available to users, and they can add functionalities for selecting data. The application represents energy consumption and offers the effect of the building shadow that is beneficial to architects and the planning body in accessing design scenarios and the regulation of rules regarding planning. The chapter primarily discusses integration practices constrained by the tools available during the experimental phase. However, it's important to note that technological advancements are rapidly evolving, suggesting that future workflows may differ significantly from those currently described.

The major gap in the proposed approach for the BIM integration in web-based GIS is the lack of connection between the design platform, the 3D model created and the 3D viewer. The changes in the 3D model in the design platform are not automatically updated in Cesium. In some cases, the components are added or deleted, and then the decomposition of the model needs to be performed manually. As it causes double work in the integration process, this issue should be solved.

BIM-GIS integration is a major topic in the field of construction and geospatial technology, and integration can enhance the current practice of data sharing in the industry. Presently, there are no official standards relating to workflows for integrating the BIM model into the GIS system, which is the main factor that makes this research and similar efforts more difficult. The major challenge is BIM's missing geographic features, which can be overcome by georeferencing the models. As none of the existing methods are fully successful, further investigation of the georeferencing techniques is required. However, the visualisation of the integrated model can be expanded by adding virtual reality (VR) in the future. The users can benefit from the VR technology as they are able to view the components of the BIM model in the provided geographic content. This section's suggestions are taken to improve the integrated BIM-GIS model's visualisation and applicability for planning a building and shadow analysis.

References

Abdeen, F. N., S. Shirowzhan and S. M. Sepasgozar (2023). "Citizen-centric digital twin development with machine learning and interfaces for maintaining urban infrastructure." *Telematics and Informatics*: 102032. https://doi.org/10.1016/j.tele.2023.102032

Abdul-Rahman, A. and M. Pilouk (2008). 2D and 3D spatial data representations. *Spatial data modelling for 3D GIS*, Springer: 25–42. https://link.springer.com/book/10.1007/978-3-540-74167-1

Akob, Z., M. Z. A. Hipni and A. A. A. A. Razak (2019). *Deployment of GIS+ BIM in the construction of Pan Borneo Highway Sarawak, Malaysia*. IOP Conference Series: Materials Science and Engineering, IOP Publishing.

Al Rifat, S. A. and W. Liu (2022). "Predicting future urban growth scenarios and potential urban flood exposure using Artificial Neural Network-Markov Chain model in Miami Metropolitan Area." *Land Use Policy* **114**: 105994.

Alexiadi, C. and C. J. F. W. W. Potsioy (2012). "How the integration of N-dimensional models (BIM) and GIS technology may offer the potential to adopt green building strategies and to achieve low cost construction." FIG Working Week, *Knowing to manage the territory, protects the environment, evaluate the cultural heritage, Rome*.

Amirebrahimi, S., A. Rajabifard, P. Mendis and T. Ngo (2016). "A framework for a microscale flood damage assessment and visualisation for a building using BIM–GIS integration." *International Journal of Digital Earth* **9**(4): 363–386.

Autodesk (2019). "Products." https://www.autodesk.com.au/.

Badea, A. and G. Badea (2018). *Considerations on open source gis software vs. proprietary gis software*. Editura Aeternitas.

Badwi, I. M., H. M. Ellaithy and H. E. Youssef (2022). "3D-GIS parametric modelling for virtual urban simulation using CityEngine." *Annals of GIS* **28**(3): 1–17.

Bansal, V. K. (2018). "Use of GIS to consider spatial aspects in construction planning process." *International Journal of Construction Management* **20**(3): 1–16.

Bansal, V. (2021). "Integrated framework of BIM and GIS applications to support building lifecycle: A move toward nD modeling." *Journal of Architectural Engineering* **27**(4): 05021009.

Basir, W., Z. Majid, U. Ujang and A. Chong (2018). "Integration of GIS and BIM techniques in construction project management-A review." *ISPRS-International Archives of the Photogrammetry, Remote Sensing and Spatial Information Sciences* **4249**: 307–316.

Bentley (2019). Bentley Software. https://www.bentley.com/software/. Access date/retrieved 1/3/2024.

Bianco, I., M. Del Giudice and M. Zerbinatti (2013). "A database for the architectural heritage recovery between Italy and Switzerland." *International Archives of the Photogrammetry, Remote Sensing and Spatial Information Sciences* **40**: 103–108.

BIMForum (2013). "Level of development specification: For building information models."

Blut, C., T. Blut and J. Blankenbach (2019). "CityGML goes mobile: Application of large 3D CityGML models on smartphones." *International Journal of Digital Earth* **12**(1): 25–42.

Borrmann, A., M. König, C. Koch and J. Beetz (2018). Building information modeling: Why? What? How? *Building information modeling*, André Borrmann, Markus König, Christian Koch, Jakob Beetz. Springer: 1–24.

Cecchini, C., A. Magrini and L. Gobbi (2019). *A 3d platform for energy data visualisation of building assets*. IOP Conference Series: Earth and Environmental Science, IOP Publishing.

Cesium (2019a). "Create, host, and fuse 3D content in the cloud." Retrieved 1/10/2019, 2019, from https://cesium.com/cesium-ion/.

Cesium (2019b). "Open source 3D mapping." https://cesium.com/cesiumjs/.

Chaturvedi, K. (2014). *Web based 3D analysis and visualisation using HTML5 and WebGL*, University of Twente Faculty of Geo-Information and Earth Observation (ITC).

Chen, Y., E. Shooraj, A. Rajabifard and S. Sabri (2018). "From IFC to 3D tiles: An integrated open-source solution for visualising BIMs on cesium." *ISPRS International Journal of Geo-Information* 7(10): 393.

Chen, J., Y. Luo, H. Zhang and W. Du (2021). "Quality evaluation of lightweight realistic 3D model based on BIM forward design." *Computer Communications* **174**: 75–80.

Chen, Y., X. Cai, J. Li, W. Zhang and Z. Liu (2022). "The values and barriers of Building Information Modeling (BIM) implementation combination evaluation in smart building energy and efficiency." *Energy Reports* **8**: 96–111.

de le Beaujardiere, J. (ed.) (2006). OpenGIS® Web Map Server Implementation Specification, Version1.3.0. Wayland, MA, Open Geospatial Consortium, 85pp. (OGC 06-042). DOI: http://dx.doi.org/10.25607/OBP-656

de Paiva, A. C. and C. de Souza Baptista (2009). Web-based GIS. *Encyclopedia of information science and technology, second edition*, Mehdi Khosrow-Pour. IGI Global: 4053–4057.

Debnath, R., C. Pettit, B. Soundararaj, S. Shirowzhan and A. S. Jayasekare (2023). "Usefulness of an urban growth model in creating scenarios for city resilience planning: An end-user perspective." *ISPRS International Journal of Geo-Information* **12**(8): 311.

Deng, Y., V. J. Gan, M. Das, J. C. Cheng and C. Anumba (2019). "Integrating 4D BIM and GIS for construction supply chain management." *Journal of Construction Engineering and Management* **145**(4): 04019016.

Dirksen, J. (2013). *Learning Three.js: The JavaScript 3D library for WebGL*, Packt Publishing Ltd.

Eastman, C., P. Teicholz, R. Sacks and K. Liston (2011). *BIM handbook: A guide to building information modeling for owners, managers, designers, engineers and contractors*, John Wiley & Sons.

Eck, D. J. (2016). *Introduction to computer graphics.* https://math.hws.edu/graphicsbook/. Retrieved 1/3/2024.

El Haje, N., J.-P. Jessel, V. Gaildrat and C. Sanza (2016). *3D cities rendering and visualisation.* (UDMV 2016), Dec 2016, Liege, Belgium, 95–100. DOI: 10.2312/udmv.20161426. hal-01787406.

Errey, J., J. Noonan and P. Brabers (2019). *Is ground modelling a waste of time?* Australasian Coasts and Ports 2019 Conference: Future Directions from 40 [Degrees] S and Beyond, Hobart, 10–13 September 2019, Engineers Australia.

ESRI (2012). "What is GIS." https://www.esri.com/library/bestpractices/what-is-gis.pdf.

ESRI (2019a). "ArcGIS." https://www.esri.com/en-us/arcgis/about-arcgis/overview.

ESRI (2019b). "Shadow impact analysis." https://solutions.arcgis.com/local-government/help/shadow-assessment/.

Fosu, R., K. Suprabhas, Z. Rathore and C. Cory (2015). *Integration of Building Information Modeling (BIM) and Geographic Information Systems (GIS)–A literature review and future needs.* Proceedings of the 32nd CIB W78 Conference, Eindhoven University of Technology, The Netherlands.

Gawley, D. and P. McKenzie (2022). "Investigating the suitability of GIS and remotely-sensed datasets for photovoltaic modelling on building rooftops." *Energy and Buildings* **265**: 112083.

Geiger, A., Benner, J. and Haefele, K.H. (2014). Generalization of 3D IFC building models. In *3D Geoinformation Science: The Selected Papers of the 3D GeoInfo 2014*, Martin

Breunig, Mulhim Al-Doori, Edgar Butwilowski, Paul V. Kuper, Joachim Benner, Karl Heinz Haefele. Springer International Publishing: 19–35.

Geoff (2017). "An open source web-based platform for integrating BIM and 3D geospatial." https://geospatial.blogs.com/geospatial/2017/12/open-source-web-based-platform-for-integrating-bim-and-geospatial.html.

Graphics, A. (2019). "Batched 3D model."

Haiying, J. (2022). "Conceptual model construction of building information management system based on BIM architecture." *Soft Computing* **26**(16): 1–7.

Hawken, S., H. Rahmat, S. M. E. Sepasgozar and K. Zhang (2021). "The sdgs, ecosystem services and cities: A network analysis of current research innovation for implementing urban sustainability." *Sustainability* **13**(24): 14057.

Hor, A., A. Jadidi and G. Sohn (2016). "BIM-GIS integrated geospatial information model using semantic web and RDF graphs." *International Archives of the Photogrammetry, Remote Sensing and Spatial Information Sciences* **3**(4): 73–79.

Ignatius, M., N. Wong, M. Martin and S. Chen (2019). *Virtual Singapore integration with energy simulation and canopy modelling for climate assessment.* IOP Conference Series: Earth and Environmental Science, IOP Publishing.

Irizarry, J., E. P. Karan and F. Jalaei (2013). "Integrating BIM and GIS to improve the visual monitoring of construction supply chain management." *Automation in Construction* **31**: 241–254.

Isikdag, U., J. Underwood and G. Aouad (2008). "An investigation into the applicability of building information models in geospatial environment in support of site selection and fire response management processes." *Advanced Engineering Informatics* **22**(4): 504–519.

Isikdag, U. and S. Zlatanova (2009). Towards defining a framework for automatic generation of buildings in CityGML using building information models. *3D geo-information sciences*. Jiyeong Lee and Sisi Zlatanova, Springer: 79–96.

Isikdag, U., S. Zlatanova and J. Underwood (2013). "A BIM-oriented model for supporting indoor navigation requirements." *Computers, Environment and Urban Systems* **41**: 112–123.

Izawa, R. and M. Koga (2019). *An extension of gltf for robot description.* 2019 19th International Conference on Control, Automation and Systems (ICCAS), IEEE.

Kaewunruen, S. and Q. Lian (2019). "Digital twin aided sustainability-based lifecycle management for railway turnout systems." *Journal of Cleaner Production* **228**: 1537–1551.

Kang, T. W. and C. H. Hong (2015). "A study on software architecture for effective BIM/GIS-based facility management data integration." *Automation in Construction* **54**: 25–38.

Kari, S., L. Lellei, A. Gyulai, A. Sik, M. M. Riedel and M. Szoboszlai (2016). "BIM to GIS and GIS to BIM." *Caadence in architecture: Back to command.* M. Szoboszlai, Budapest University of Technology: 67–72.

Karimi, K., N. G. Dickson and F. Hamze (2010). "A performance comparison of CUDA and OpenCL." *arXiv preprint arXiv:1005.2581.*

Khemlani, L. J. W. p. (2004). "Autodesk Revit: Implementation in practice." Autodesk.

King, M. (2019). "Review of getting to know web GIS." *Cartographic Perspectives* **92**: 100–102. DOI: 10.14714/CP92.1537

Knoth, L., M. Mittlböck, B. Vockner, M. Andorfer and C. Atzl (2019). "Buildings in GI: How to deal with building models in the GIS domain." *Transactions in GIS* **23**(3): 435–449.

Lacaze, B., J. Dudek and J. Picard (2018). "Grass GIS software with QGIS." *QGIS and Generic Tools* **1**: 67–106.

Lapierre, A. and P. Cote (2007). *Using Open Web Services for urban data management: A testbed resulting from an OGC initiative for offering standard CAD/GIS/BIM services*. Urban and Regional Data Management. Annual Symposium of the Urban Data Management Society, Taylor & Francis, London.

Lee, N., T. Salama and G. Wang (2014). Building information modeling for quality management in infrastructure construction projects. *Computing in Civil and Building Engineering* **2014**: 65–72.

Liu, F., H. Zhang, Y. Hu, X. Guo, Z. Zhu, J. Jia and H. Zhu (2019). *Cesium based lightweight WebBIM technology for smart city visualization management*. International Conference on Inforatmion Technology in Geo-Engineering, Springer.

Ma, Z. and Ren, Y. (2017) "Integrated application of BIM and GIS: an overview." *Procedia Engineering* **196**: 1072–1079. DOI: 10.1016/j.proeng.2017.08.064.

Makkonen, S. (2016). Semantic 3D modelling for infrastructure asset management (Master's thesis). https://aaltodoc.aalto.fi/items/ae12fb52-0ba4-45ba-b82d-b6bf6ab86256

Manolova, M. (2018). "Integration of 3D BIM models in a web GIS for life cycle asset management." https://repository.tudelft.nl/islandora/object/uuid:2a2523d1-c751-44e2-8129-676ef1eec379

Masoumi, H., S. Shirowzhan, P. Eskandarpour and C. J. Pettit (2023). "City digital twins: Their maturity level and differentiation from 3D city models." *Big Earth Data* **7**(1): 1–36.

Matos, R., H. Rodrigues, A. Costa and F. Rodrigues (2022). "Building condition indicators analysis for BIM-FM integration." *Archives of Computational Methods in Engineering* **29**(6): 3919–3942.

Matsuda, K. and R. Lea (2013). *WebGL programming guide: Interactive 3D graphics programming with WebGL*, Addison-Wesley.

Maurya, S. P., A. Ohri and S. Mishra (2015). Open source GIS: A review. In *Proceedings of National Conference on Open Source GIS: Opportunities and Challenges*. Banaras Hindu University: 150–155.

Mete, M. O., D. Guler and T. J. S. Ü. M. Yomralioglu, Bilim ve Teknoloji Dergisi (2018). "Development of 3D web GIS application with open source library." Selçuk Üniversitesi Mühendislik, Bilim Ve Teknoloji Dergisi, **6**: 818–824.

Mignard, C. and C. Nicolle (2014). "Merging BIM and GIS using ontologies application to urban facility management in ACTIVe3D." *Computers in Industry* **65**(9): 1276–1290.

Mozilla (2019). "How the web works."

National Research Foundation (2020). "Virtual Singapore." Retrieved 22/05/2020, 2020, from https://www.nrf.gov.sg/programmes/virtual-singapore.

Park, S.-H. and E. Kim (2016). "Middleware for translating urban GIS information for building a design society via general BIM tools." *Journal of Asian Architecture and Building Engineering* **15**(3): 447–454.

Pauwels, P., S. Zhang and Y.-C. Lee (2017). "Semantic web technologies in AEC industry: A literature overview." *Automation in Construction* **73**: 145–165.

Prandi, F., F. Devigili, M. Soave, U. Di Staso and R. De Amicis (2015). "3D web visualisation of huge CityGML models." *International Archives of the Photogrammetry, Remote Sensing & Spatial Information Sciences* **40**: 601–605.

Rafiee, A., E. Dias, S. Fruijtier and H. Scholten (2014). "From BIM to geo-analysis: view coverage and shadow analysis by BIM/GIS integration." *Procedia Environmental Sciences* **22**: 397–402.

Reinhardt, J. and Bedrick, J. (2013). Level of development specification for building information models. In *BIM Forum*. Overall co-chairs Jan Reinhardt, Adept Project Delivery Jim Bedrick, FAIA, AEC Process Engineering. https://bimforum.org/wp-content/uploads/2022/06/BIMForum_LOD_2013_reprint.pdf.

Sayar, A., M. Pierce and G. Fox (2005). *Developing GIS visualisation web services for geophysical applications*. ISPRS 2005 Spatial Data Mining Workshop, Ankara, Turkey. International Society of Photogrammetry and Remote Sensing.

Schilling, A., J. Bolling and C. Nagel (2016). *Using glTF for streaming CityGML 3D city models*. Proceedings of the 21st International Conference on Web3D Technology, ACM.

Schueren, M. (2017). "Why 3D GIS?" https://www.linkedin.com/pulse/why-3d-gis-madeline-schueren/.

Sepasgozar, S. M. E., A. M. Costin, R. Karimi, S. Shirowzhan, E. Abbasian and J. Li (2022). "BIM and digital tools for state-of-the-art construction cost management." *Buildings* **12**(4): 396.

Sepasgozar, S. M., A. A. Khan, K. Smith, J. G. Romero, X. Shen, S. Shirowzhan, H. Li and F. Tahmasebinia (2023). "BIM and digital twin for developing convergence technologies as future of digital construction." *Buildings* **13**(2): 441.

Sergi, D. M. and Li, J. (2014). "Applications of GIS-enhanced networks of engineering information. *Applied Mechanics and Materials*, **444**: 1672–1679.

Services, S. (2020). "Digital twin." Retrieved 22/05/2020, 2020, from https://www.spatial.nsw.gov.au/what_we_do/projects/digital_twin.

Shirowzhan, S., W. Tan and S. M. E. Sepasgozar (2020). "Digital twin and CyberGIS for improving connectivity and measuring the impact of infrastructure construction planning in smart cities." *ISPRS International Journal of Geo-Information* **9**(4): 240.

Soward, E. and J. Li (2021). "ArcGIS Urban: An application for plan assessment." *Computational Urban Science* **1**(1): 1–10.

Stähl, L. (2017). "Cesium vs. ArcGIS API for JavaScript." http://www.ika.ethz.ch/studium/cartography_lab/2017_staehli_report.pdf.

Steiniger, S., R. Weibel and B. Warf (2010). "GIS software: A description in 1000 words." https://www.zora.uzh.ch/id/eprint/41354/8/Steiniger_Weibel_GIS_Software_2010.pdf

Tao, W. (2013). "Interdisciplinary urban GIS for smart cities: Advancements and opportunities." *Geospatial Information Science* **16**(1): 25–34.

TechwireAsia (2020). "Here's how the world uses digital twins to solidify smart city development." Retrieved 28/05/2020, 2020, from https://techwireasia.com/02/2020/heres-how-the-world-uses-digital-twins-to-solidify-smart-city-development/.

Tekla (2019). "Software solution."

Terres de Lima, L., S. Fernández-Fernández, J. F. Gonçalves, L. Magalhães Filho and C. Bernardes (2021). "Development of tools for coastal management in Google Earth engine: Uncertainty bathtub model and Bruun rule." *Remote Sensing* **13**(8): 1424.

Todorović, M., S. Živković, L. J. T. a. a. r. f. t. p. Haznadarević and l. a. o. t. papers. (2018). "Managing projects in the field of enviroment and life quality." XIII. Znanstveno-stručna konferencija s međunarodnim sudjelovanjem „Manadžment i sigurnost", Čakovec, Hrvatska, Zbornik radova, str. 295–308. ISBN 978-953-55241-6-8 UDK 005:331.4(063) UDK/UDC, 005.73:331.45.

Tsou, M.-H. and J. Smith (2011). "Free and open source software for GIS education."

Twinmotion (2019). https://www.twinmotion.com/en-US/download

van Rees, E. (2014). "FME 2014 release." *GeoInformatics* **17**(2): 10.

Vandezande, J., P. Read and E. Krygiel (2011). *Mastering Autodesk Revit architecture 2012*, John Wiley & Sons.

Vretanos, P. A. J. O. G. C. S. (2005). "Web feature service implementation specification." **1325**: 04–094. Version 1.1. 0. https://repository.oceanbestpractices.org/handle/11329/1128.

Wang, H., Y. Pan and X. Luo (2019a). "Integration of BIM and GIS in sustainable built environment: A review and bibliometric analysis." *Automation in Construction* **103**: 41–52.

Wang, M., Y. Deng, J. Won and J. C. Cheng (2019b). "An integrated underground utility management and decision support based on BIM and GIS." *Automation in Construction* **107**: 102931.

Wang, Q. and Y. Li (2022). "Research on simulation of distribution network engineering scene based on 3D GIS technology." *Wireless Communications and Mobile Computing* **2022**: 1–9.

Wen, Q.-J., Z.-J. Ren, H. Lu and J.-F. Wu (2021). "The progress and trend of BIM research: A bibliometrics-based visualisation analysis." *Automation in Construction* **124**: 103558.

Whiteside, A. and J. Evans (2007). *OpenGIS® Web Coverage (WCS) implementation specification, version 1.1. 0*, Open Geospatial Consortium Inc.

Xu, S., S. Shirowzhan and S. M. Sepasgozar (2023). "Urban waste management and prediction through socio-economic values and visualising the spatiotemporal relationship on an advanced GIS-based dashboard." *Sustainability* **15**(16): 12208.

Yoders, J. J. R. M. (2013). "Collaboration: How architects and engineers can work together with BIM in the cloud." **25**: 2014.

Zhao, L., J. Mbachu and Z. Liu (2022). "Developing an integrated BIM+ GIS web-based platform for a mega construction project." *KSCE Journal of Civil Engineering* **26**(4): 1505–1521.

8 BIM and GIS values and acceptance criteria in government-funded infrastructure projects

A case study of Australia

Lee Butler, Sara Shirowzhan and Samad M. E. Sepasgozar

BIM and GIS are known as key tools for digital transformation in the architecture, engineering, and construction (AEC) industry. Governments promote the use of these modelling systems to enhance collaborative solutions, review documentation, and streamline the construction project cycle. This study investigates practitioners' intentions to utilise these technologies due to a lack of investigation into government-funded infrastructure projects.

This research investigated the key barriers to the slow adoption rates of both BIM and GIS in infrastructure projects and examined what strategies can be developed to improve adoption. The research conducted a comparative review of the status of overseas practices through an extensive literature review process.

The research suggests an industry that lacks consistency in government-funded projects in terms of BIM and GIS implementation due to the diversity of projects, fragmented documentation, and differences in the capability of managing information systems among all players in the industry. The chapter also suggests that a government drive to implement BIM or GIS is required to alleviate the disarray overshadowing the technology. The use of BIM and GIS depends on the value and usefulness of the designed system. The findings also identified how technical, managerial, and cost obstacles affect the project benefit across three stages of the project.

Primary data was consolidated from a limited sample of Australian construction industry practitioners through an online survey questionnaire. The primary data research was designed using a top-down approach. Theoretical variables were questioned at a macro level and later broken down into specific groups. Secondary data was examined through the literature review and helped to form the framework of a roadmap for BIM implementation. The data obtained was examined through mixed-method analysis. Despite the number of Australian infrastructure projects being recently approved, past projects have suffered from poor performance, which has led to cost and project overruns and, in some cases, project failure.

DOI: 10.1201/9781003507000-8

8.1 Introduction

The revolutionary development of digital technologies has been welcomed in AEC in the past two decades. The initial concept of building information modelling (BIM) was first announced as far back as 1962. Eastman (2008), regarded as the "father" of BIM, claimed the method would reduce design costs for issues related to drafting and document transmittal. However, inconsistency and inaccuracy of drawings still play a significant part in the project, leading to cost overruns in the construction industry and disputes amongst shareholders (Eastman 2008). Ten years later, "Australia 2030: Prosperity through Innovation" (Department of Industry, Science, Energy and Resources 2018) announced its plan to lead the global innovation race, a race that is likely to see significant changes to the AEC industries' processes, marketing techniques, and huge organisational benefits. More recently, a survey conducted of 63 nations ranked Australia at 17 (Institute of Management Development 2020).

Whilst there are many studies that highlight the three common factors for BIM's slow adoption rates, namely technology, process, and people (Khosrowshahi and Arayici 2012), there is a dearth of investigations for both BIM and GIS acceptance in infrastructure projects in particular, those funded by the government. GIS is one of the critical systems required for infrastructure projects, but it is mainly ignored in many investigations in the construction literature. This research investigates what other factors may impact the implementation traction of these information systems and develops a theoretical model including critical constructs.

The Australian Commonwealth Government has supported state governments with the suite of infrastructure projects that are currently being undertaken nationally. The motivation behind the research began with a series of organisational changes within the construction industry. The resistance to changing conventionally used systems without fully understanding the benefits of BIM has held the Australian construction industry back from further progress. The global construction industry across many developing regions has taken great interest in digitalisation, with the key driver for BIM implementation in Australia being industry-led. Sobeit, Australia, has witnessed slow adoption rates, particularly within SMOs. The uncertainty and lack of knowledge about BIM technology, return on investment, and lack of government incentives have caused the SMOs to oppose the technology. Despite a partial mandate within commonwealth projects exceeding \$50 million and \$30 million on NSW Health Infrastructure projects, there has been no national mandate, Australian standard, or a clear roadmap for the future of BIM. There is an apparent need for immediate government action to address these issues.

To maximise the strengths of qualitative and quantitative analysis, a mixed-methods approach was adopted to analyse the barriers to BIM adoption and to examine the hypothesis of BIM's slow adoption rates. The method proposed is highly relevant to the aims and objectives of this chapter. The chapter used a critical literature review process that analysed key research chapters from a period within the past ten years on subjects such as construction, BIM impacts, and the Australian Government, together with a questionnaire survey.

While the fundamental goal of many BIM-associated reviews undertaken was to outline the past and current challenges to BIM adoption, there is limited research on BIM, specifically on Australian Government-funded projects. Many researchers would say we are in the middle of a digital revolution; in fact, some are calling it the fourth industrial revolution. We are moving faster today than we ever have before (Leonhard 2016). Leonhard (2016) argues we, and the world's technology, are in one of the most transformational times experienced in human history. Examples include the introduction of the early versions of the iPhone in 2007 to self-driving cars and computers that can learn, think, predict, and decide (Saranya and Subhashini 2023).

This research looks at the forecast for BIM, how the construction industry is adapting to digitalisation and new technologies, and, therefore, how the AEC industry will react to emerging new challenges (The University of British Columbia 2021; Sepasgozar, Khan et al. 2023).

The term "digital economy" refers to the series of economic and social activities supported by information and communication technologies (Cardeira 2019; Luo, Yimamu et al. 2023). However, the use of digital technologies to change a business model with the objective of generating new revenue and value is known as "digitalization". The geographic information system (GIS) is an advanced technology system that connects and manages data to a map and has the ability to capture and analyse geographic and spatial data (Esri 2021; Visner, Shirowzhan et al. 2021; Zhu, Shirowzhan et al. 2022). The study explored the relationship of BIM-GIS integration and where it was best utilised.

8.2 Overseas context

The United States has been recognised as an impressive motivator for digital growth, which is driven by its research and development agility and its optimistic approach to embracing new digital technologies (Institute of Management Development 2020). In fact, 72% of all US construction firms are using BIM (Griffey 2019).

The benefits of BIM have been experienced by many overseas countries, yet Australia has been slow to adapt (Oyewole and Dada 2019). Finland has taken the lead in the implementation of BIM, followed by the United Kingdom and North America (Johnston 2008). The UK announced its future strategy in 2011. It requires all government tendered projects to be at BIM Level 2, i.e., 4D/5D full collaboration, and BIM Level 3 on all centrally procured projects by 2016 (United-BIM Inc. 2020).

BIM's slow uptake has been narrowed down to three factors: technology, process, and (Khosrowshahi and Arayici 2012). This chapter seeks to answer why very slow progressives occur and how momentum can be gained in BIM implementation in the Australian Government construction sector; once these questions are answered, strategies are recommended for developing the business model for the Australian construction sector.

Australian organisations traditionally use 2D AutoCAD drawings to service their projects. Gerrard, Zuo et al. (2010) argue that the root cause for slow adoption is the level of BIM expertise, lack of awareness, and resistance to change. However, changing an established company's culture and people's attitude towards BIM won't be the silver bullet to BIM's overnight success but rather pave the way to a successful future.

The output of a report conducted on 63 nations' participation concluded that Australia was placed number 13 in the digital competitiveness ranking (Bris, Cabolis et al. 2017) and more recently ranked 17 (Institute of Management Development 2020). In an environment where data processing is still done manually, Australia needs a dynamic shift towards digitalisation. To compete with overseas construction models, Australia needs to capitalise on newly available technologies to improve its infrastructure project lifecycle. In the past decade, Australia's appetite to build larger and more complex projects has grown (Infrastructure Australia 2019). A focus on methods to improve project delivery by using new technology and digitalisation has been BIM-based (Infrastructure Australia 2016).

8.3 Australia as a research context

Following the 2016 Australian Government announcement regarding the implementation of BIM (Griffey 2019), a government-driven approach has been developed to support BIM in future Australian Government-funded projects. Based on the UK BIM Task up model, Infrastructure Australia will lead a task force to coordinate and develop BIM's national implementation. The government has decided to make BIM a compulsory requirement for government-funded projects exceeding $50 million. The policy stipulates the early adoption of BIM by a project's affiliated agencies during the tender stages. The clients of building projects are the major beneficiary of BIM implementation (Yang and Chou 2018). With these conditions, the Australian Government is likely to receive significant benefits from BIM implementation.

NSW has multiple mega projects. Sydney Metro, Australia's biggest-ever rail project, Northwest Rail Link, and West Connex are just three of the infrastructure projects being delivered. Furthermore, NSW has committed to investing $108.5 billion into newly selected infrastructure projects over a four-year period (Shepherd 2021). The Northwest Rail Link and Barangaroo development (including Wynyard Walk) are good examples of BIM utilisation. The web-based project management system was developed and tailored to control and evaluate data to efficiently manage the entire lifecycle of construction projects, which in this case includes the construction and operation of the rail project.

In 2013, NSW Health Infrastructure introduced the first step to mandate BIM on projects exceeding $30 million; however, the national government decided not to impose BIM on public projects. The evidence is clear of BIM's extensive capabilities, i.e., enhancing performance, improving design communication, early clash detection, and improving project communication amongst stakeholders. However, some projects have not fully embraced the adoption of BIM, and the reports show the same issue in some developing and developed countries (Oyewole and Dada 2019). Gerrard, Zuo et al. (2010) argue that the root cause for slow adoption is the level of BIM expertise, lack of awareness, and resistance to change.

When discussing implementation rates, Australia is trailing the industry-leading BIM pilots (Griffey 2019). However, the industry is witnessing pivotal change. The international implementation of BIM is maturing (Queensland Government Enterprise Architecture 2020). The Queensland Government included BIM in their programs, including local government and planning policies and principal document in 2017. The Queensland Department of Transport and Main Roads has some guides for using BIM on infrastructure projects. In the 2015/2016 Victorian Government budget, the strategies were announced to intensify their efforts in BIM research and adoption. Selected projects would participate in BIM pilot surveys to assist with Victoria's future development (Institute of Public Works Engineering Australasia 2019a). In Western Australia, the State Government has started to use BIM on some of its high-profile infrastructure projects, mostly at the procurement stages.

Although there has been some evidence of progression, the Australian model lacks specific Integrated Project Delivery (IPD) contracts. IPD contract models usually consist of: (1) The client, (2) Head contractor, (3) Designer. IPDs are a form of collaborative contract. Therefore, risk and rewards are shared amongst the parties involved through a profit/incentive pool that is determined by measurable project outcomes (Seyfarth 2019).

The Australian construction industry has yet to experience BIM's full capabilities. Although the current benefits have been analysed, they conclude that definitive benefits have not been fully capitalised by industry stakeholders (Ghaffarianhoseini, Tookey et al. 2017). Further research is required to fully understand the system's benefits.

However, cases of project and stakeholder communication inefficiencies are common. One must ask why these issues haven't been addressed. Are the keys to the barriers of implementation hidden inside projects? This chapter will attempt to answer this problem. The purpose of the method behind the final chapter is to analyse the barriers to Australian Government sector implementation, including the rationale behind industry resistance slow adoption the strategies that could be developed to improve the Australian BIM adoption model.

8.4 Examples from past projects

Past government-funded projects procured through design and construct (D&C) packages resulted in main contractors taking on the greater share of risks (Australian Government 2015). The NSW auditor illustrates how delays and cost overruns are common throughout this sector. A fragmented exchange of information in a project environment leads to information being lost due to inefficient and ineffective communication among project stakeholders.

Previous relationships between contractor and client on major projects have resulted in legal disputes costing as much as $1 billion, to cite one case involving the NSW Government and ALTRAC (Kemp 2019). Australia has taken a much more cautious approach. The NSW State Infrastructure Strategy 2018–2038 (NSW Government 2021) outlines a 20-year development framework that sets out the

Government's priorities and creates a pipeline of future investment in major projects. Future projects will experience a more collaborative and open book policy with a drive to build relationships with key players in the construction industry. The approach encourages efficient practices, innovation, communication, and collaboration. A key part of the strategy was collaborative contracting. This involved the segregation of larger contracts, allowing not only top-tier contractors but also small- to medium-sized organisations (SMO) to be involved in the planning and delivery of infrastructure projects. This plan would encourage innovation and competitive contracting while expanding the contractor participants for a future project.

There is a significant gap in the Australian construction industry. Although BIM is a powerful tool in theory, partly due to its immaturity, its practical application performance to reduce information asymmetry is yet to be confirmed (Forsythe, Sankaran et al. 2015).

The chapter will review the feedback from industry professionals and discuss areas where BIM is lagging and where improvements can be made in its Australian implementation. There has been a fundamental shift in the construction industry towards digitalisation. More recently, according to a report completed by Ribeirinho, Mischke et al. (2020), the COVID-19 pandemic has accelerated organisations' digital transformation. Over half of the participants reported that they had already raised their level of digitisation investment. Advanced technologies are rapidly changing our society, industry, economy, and our everyday lives (Leonhard 2016). New technologies and increased rates of digitalisation have put further pressure on the construction industry to change its ways.

8.5 Development of BIM

The construction industry and its supporting ecosystem have been underperforming in recent years. Uncertain market factors and competitive industry dynamics have made attempts at change mostly difficult. Research completed in the United Kingdom (UK) by Khosrowshahi and Arayici (2012) concluded there was a need to gain momentum in BIM implementation. Khosrowshahi and Arayici (2012) also seek to answer the question of why the UK is experiencing a fragmented uptake of BIM. Further to reviewing the UK study, this chapter explores ways to develop a roadmap that is specific to the Australian implementation and makes a basis for recommending strategies for Australian BIM adoption. Whilst the uptake of BIM remains relatively slow, the implementation of BIM should not be forced (Forsythe 2014). Instead, a more planned approach in stages with the necessary training and teaching requirements would have far greater outcomes. The Australian Government has decided not to enforce the use of BIM on government-funded projects and prefers to take a gradual approach. Its current requirements are BIM to be at LOD 500 – As-Built on all commonwealth-funded projects exceeding $50 million. The state Government will follow similar principles.

The key factor of successful construction projects is establishing the correct balance between contractual and relationship issues with healthy partnership delivery. BIM's previous Australian studies found a lack of symmetry (Forsythe, Sankaran

Table 8.1 BIM dissemination cause/effects

Causes	*Effects*
Transient teams	Construction teams segregate once the project is delivered
Stakeholders (AEC) loyalty	Fragmented information on future projects
Retirement of employees	Loss of knowledge and experience
Loss of documentation	Loss of knowledge and records
Information embedded in filed documents	Disintegration of knowledge and events
Rejected value engineering proposals, and change orders	Scattered information across the company and stakeholders

et al. 2015) an issue which commonly appears in contracting at project procurement stages, when an an unevenness is discerned between the two contracting parties. Forsythe, Sankaran et al. (2015) identified the three main causes: adverse selection, moral hazard, and hold-up. They explored the possibility that BIM could alleviate the imbalance in information. The research suggested that contractual relationships between client, designer, and contractor required further development. Secondly, a major push is required by industry peak bodies and educational institutions to train and develop a workforce that is "BIM ready," a workforce that has the capacity to innovate construction industry work practices. Lastly, the limitations were set, and further studies were recommended across Australia, including studies on supply chains.

Similar studies conducted by Deshpande, Azhar et al. (2014), approached this topic in a different context. The study argues the need for systematic knowledge management to encourage continuous improvements, share the project's wealth of knowledge, help transfer new knowledge to innovative practices, reduce reworks and faster responses from customers, and improve overall project performance. Each construction project forms a new set of complexities and issues that require hands-on problem-solving techniques throughout the project lifecycle. The wealth of knowledge gained during this stage is, according to the author, the AEC firm's most important asset. Furthermore, Deshpande, Azhar et al. (2014) also found effective management of knowledge to be challenging for AEC firms. Some key reasons are proposed in Table 8.1.

The table outlines a range of challenging issues related to knowledge and information dissemination. Deshpande, Azhar et al. (2014) also describe BIM impact as unique and revolutionising the way AEC projects are delivered. BIM may offer a possible solution to the issues identified in Table 8.1. A platform that can integrate dispersed information and visually represent the information, improving collaboration and communication of construction stakeholders.

8.6 The implementation of BIM

Australia's BIM implementation is still at the infancy level. The maturity of digital tools has improved the communication between clients, consultants, firms, and governments. Economies worldwide have seen the benefits and importance of digital activities. This innovative shift has increased the demand for statistical insights into digital activities in Australia for analytical and policy purposes (Australian Bureau of Statistics 2019). The construction industry is a major player in the Australian economy and accommodated over 950,000 employees in 2013 (Australian Bureau of Statistics 2013). By 2023, it has been estimated that it will employ an additional 318,800 (Back to Basics 2019).

During the period 2011–2012, the public sector accounted for $305.5b of income revenue.

A survey conducted by Elmualim and Gilder (2014) found that client architects and design managers had the most influence on bringing innovation into a project. Most respondents to the survey concluded that greater usage of BIM would result in significant improvement to overall performance. However, the reasons raised for not implementing BIM, ranged from software/hardware risk liability issues to benefits that did not outweigh the onboarding costs, and firms were hesitant to initiate new workflows or re-train their existing workforce. Moreover, almost half of the respondents did not know why their organisations hadn't adopted BIM. Elmualim and Gilder (2014) summarise the three key challenges of the implementation: training staff on the BIM process; implementing new processes and workflows, and understanding BIM implementation processes sufficiently.

An alternative study conducted by Bridge and Carnemolla (2014) explored BIM's involvement in social inclusion and looked at life without BIM. It found a lack of supporting design tools for architects and designers. It also suggests that life without BIM would lead to environments that merely comply with a minimum "deemed to satisfy" building codes, resulting in reworks that required additional funding taken from taxpayer payments.

Previous Australian research concluded that businesses that invest in digital technologies result in higher productivity (Australian Government 2018) so there must be key implementation challenges that remain *i.e.*, a slow-to-innovate industry, the effort to stay competitive, and time and cost. To overcome these challenges, a new approach is required. Despite the gradual move led by Infrastructure Australia, the government needs to champion this move and look at ways to support and encourage the adoption rate and further the national implementation rate.

8.7 Open and closed BIM and lean construction

Closed BIM (CB) has been used on projects that require selected stakeholders to access the same software whilst constructing the project. Despite this method allowing some form of participant collaboration, it does not allow open integration.

However, open BIM (OB) is an improvement in terms of accessibility, usability, management, and sustainability of digital data in the built asset industry (Australasia 2012). Central Europe and the Nordics have already seen the benefits of OB and implemented OB models on some of their rail infrastructure projects. It is based on open standards and workflows. The key principles that OB recognises are interoperability, openness, reliability, collaboration, flexibility, and sustainability.

OB greatly enhances the useability of BIM. It enables the user to share their data with any BIM-compatible software. There are clear benefits to adopting processes related to OB in the construction industry.

Lean construction (LC) aims to improve productivity, eliminate waste, cut costs, and reduce construction duration, resulting in safer and more effective projects. Integrating LC with BIM on projects results in full stakeholder collaboration, increased profits for all involved, supply chain integration, and continuous improvements (ZIGURAT Global Institute of Technology 2018; Sepasgozar, Hui et al. 2021). The Queensland Government (QG) describes how its departments are about to embark on a series of pilots that use these new technologies and initiatives (Queensland Government 2021). The plan outlines an implementation strategy focusing on developing BIM and digital engineering in a common data environment. The objective is to deliver infrastructure faster, at less cost, and safer than with previous projects. The QG recognises that industry has been challenged to accept that we cannot maintain the same processes and actions for ever.

8.8 Digital engineering framework

Although it has been discussed in many studies and business models, the main drivers for the implementation of project digitalisation are people, processes, and technology. However, many mid-level executives and experts in the field do not speak nor process data efficiently (Logan 2018). This study argues that without clear processes and well-governed data, the success rates are minimised (Hamid 2021).

Research undertaken on BIMs' ability to transform cities (Fischer 2016) found a need to step aside from a single project-focused mindset and approach rapid learning across multiple projects. A consolidated valuable information source (i.e. data) will enable BIM users to expeditiously predict the performance of projects. Not only is information gathered through fieldwork experience, but equally important is digital experience. Fischer (2016) suggested that access to good-quality data is indispensable to optimise selected factors. Fischer (2016) concludes by mentioning that capturing expertise and data together enables the collected data to be modelled and visualised, which assists us with the essential knowledge to achieve the vision of digital cities.

Logan (2018) describes the distance between the designers (creators) and the users (consumers) as being worlds apart. The author compared a veteran versus a rookie, i.e., a person with a 30-plus-year career as opposed to a brand-new PhD graduate in data science. Despite being diverse, these professionals have no shared common language at two ends of the scale, leading to a fundamental communication challenge when utilising data-driven and analytical solutions. Logan (2018) concludes data must be the focal point of digital society.

Despite BIM being the catalyst in the evolution of the construction sector, it has also led analysts and industry professionals to refocus on the way they conduct information management (Vaux 2021). Vaux (2021) explains a five-step process toward smart cities, with BIM being the starting point. The next step is the creation of a digital ecosystem linking multiple data sets together with common data. The digital engineering approach is paramount to unlocking future innovations such as Digital Twins and smart infrastructure. Vaux (2021) also emphasises that instead of focusing on the common three-pillar organisational model: people, process, and technology, the focal point needs to be data-centric.

If data is the new core capability of digital transformation, then prioritising data as a language that all levels of employees must familiarise themselves with is crucial. Not only does data play a key role in the initial stages of the project *i.e.*, service, network, and project planning, but it is equally important throughout the project management phases through to handover, asset management, and asset renewal. At the core of the stages listed, a data management business needs to be maintained to be effective.

The current state of the market is in disarray with a wide range of disconnected data and fractured environments (Vaux 2021). The United Kingdom (UK) has taken a giant leap forward, clearly set out in the "Digital Roads 2025 vision" (Highways England 2014). In 2019, they recognised that data is an asset itself and have now backed data management with strong governance. Moreover, any data created is infused into the central lifecycle so it can be repeatedly reused and open opportunities to return value to the organisations involved. The UK is embarking on a digital technology revolution that will transform its national highways, including electric, automated, and connected vehicles, and significantly improve the future road network (Dyson 2021). The strategy has taken a "no stone unturned" approach to the "once-in-a-century" (Marsh 2021) transformation of how UK roads are designed, built, operated, and maintained. The strategy has been forecasted to create greener, safer, more inclusive, and reliable mobility by harnessing technology embedded into an intelligent system. Digital engineering is working to achieve a common objective to create a standardised code for each individual type of dataset to achieve its ultimate goal: a digital ecosystem of linked databases.

8.9 Research method

This research used a mixed-method approach to analyse primary and secondary data sources. A comprehensive literature review was undertaken in the first part of the study to understand the status and hypothesis surrounding the uptake, barriers, and benefits of BIM in the Australian construction Industry.

The second part of the study relied on key industry players as its source of primary data collection. A questionnaire survey was undertaken to collect the primary data, which formed the primary data source and obtained a good understanding of market conditions (Sage Research Methods 2012). Industry practitioners were invited to participate, and once they agreed, they were eligible to participate in the survey.

A thematic analysis was utilised to analyse data and scrutinise the findings of the questionnaire survey at different levels. Additionally, the thesis analysed data through one-way testing, reliability, and factor analysis. There are very few literature publications available on BIM within commonwealth projects. Through qualitative analysis and thematic analysis, conducted with the results of the survey questionnaire, this research attempts to address the gap related to the hypotheses surrounding this topic.

An in-depth review was performed to provide transparency across multiple relevant literature (Elghaish, Matarneh et al. 2021). Utilising a variety of online search engines, namely, Scopus, UNSW library and Google Scholar, the research targeted literature classifications dating from 2011 to 2021. Key search words used included "BIM adoption," "Barriers of implementation," "Australian Government," "Collaboration," "Innovation," "Roadmap," "Dissemination," "BIM GIS fusion," and "Integration." The literature was downloaded and categorised into selective groups with sub-group categories.

8.9.1 Research design

The mixed-methods approach (qualitative and quantitative) was first utilised to conduct the literature review and compile the questionnaire survey. By examining the literature, repeated problems and benefits were noted. A more detailed analysis of the issues and benefits found a common frequency occurring.

This research was conducted through a survey of experts in the construction industry based on an approved ethics protocol of HC210704. The practitioners were carefully selected to ensure an accurate analysis was completed. The research found many ex-AutoCAD (CAD) users were transitioned into BIM expert roles but did not fully understand the capabilities of BIM. To maintain response accuracy, the survey aimed at high-level experts, including BIM co-ordinators, modellers, detailers, delivery project managers, commercial managers, digital engineers, general managers, executive directors, global digital transformation leads, senior design engineers, design managers, planning and project controller, and specialists. The questionnaire survey narrowed the groups down to participants with experience in infrastructure projects, such as tunnel engineers, metro, rail and infrastructure project managers, road and highway infrastructure engineers, rolling stock project engineers, transport design leaders, and transport planners.

The categories of questions were selected based on the extensive research conducted in the literature review. The experience of the industry participants ranged from 4 to 25 years.

The survey results were analysed using various tests and statistical analysis. To examine the precision, effectiveness, and suitability of the questions, a pilot survey was conducted with 15 random participants. The feedback generated from the pilot survey was not used in the final analysis; however, it assisted the researcher in refining and developing the final version for distribution.

The data collection commenced with an online survey where a grand total of 200 survey invitations were targeted and distributed through online professional social networks and emails to BIM-GIS industry experts. The purpose of the questionnaire was to obtain current primary information on the position of the Australian industry and to provide a clearer vision of the proposition surrounding the Australian BIM resistance.

8.9.2 Measuring sampling

Data was collected from respondents using a mixed-methods approach survey questionnaire based on qualitative and qualitative analysis to examine the perception of BIM users and non-users in the context of its implementation and status within the Australian construction industry. This method was selected to utilise the maximum amount of data from the respondents. The analysis also tested the validity and reliability of relationships between key industry practitioners. Therefore, the survey participants targeted the key industry practitioners listed in the following paragraphs during recruitment. Of the 200 invitations distributed, 55 responses were collected, with a valid response rate of 27%.

This survey used Qualtrics, an online administered questionnaire, to analyse the respondent data. The software allowed the researcher to store, manage, and classify data (via coding), analyse data, and represent textual data results in a basic graph/chart format.

8.9.3 Recruitment and data collection

Participants from Australia were invited to complete the survey, and the UNSW template invitation format, which adhered to the human ethics requirements, was utilised. The survey was voluntary and remained anonymous throughout the recruitment and data collection phase.

The data collection commenced with an online survey where a grand total of over 200 questionnaire survey invitations were distributed through online professional social networks and emails to BIM/GIS industry experts, including BIM Manager/Co-ordinator, Construction Interface Manager, BIM Specialist, Consultants – Specialising in using BIM, GIS & Digital Twins, Project/Site Engineer – Vast experienced in civil/infrastructure projects, Structural Engineering, BIM Manager, Designers/Architects, BIM/Modeler, Draftsmen, Service Managers, Estimators, Building Managers, Tunnel Engineers, Digital Delivery Manager, Autodesk Civil and Infrastructure Support.

The questionnaire was designed to extract information from four main categories and a limited number of subcategories. In the first section, the survey consisted of personal and general questions, including expertise, industry experience, academic qualifications, location, geographical origin of the organisation, BIM/GIS awareness, and whether the organisation represented by the participant was a BIM/GIS user or non-user.

The next section looked at technical questions related to software and integration with other digital platforms. Three questions related to BIM-GIS integration technology.

Many previous articles have described the benefits of BIM from the cradle to the grave of a project life cycle, including project initiation, design, evaluation, construction, operation, and demolition (Barlish and Sullivan 2012; Dowsett and Harty 2013; Lu, Peng et al. 2013; Terreno, Anumba et al. 2015). However, none have emphasised the benefits of BIM concerning Australian infrastructure projects.

The questions were developed to measure and calibrate the current theoretical level of BIM in Australia.

The survey looked at implementation, technology, and training. The survey strives to bring attention to the extent of the issues relating to the value of BIM technologies, the level of BIM being utilised, the types of training, and company policy towards BIM and/or GIS.

Finally, the survey looked at obstacles and open questions. The obstacle section of the survey was a key initiative and an important topic within the research. The knowledge gained from the data could be a fundamental factor as an enabler of BIM implementation.

The rationale for including open-ended questions in the final stages was to provide the participants with an environment where they could generate more in-depth data or stories. The data was then extracted by qualitative analysis that related to the sample population as a tool to capture outstanding issues. The survey questions were designed to extract data about BIM user needs and current adoption, implementation, and impact issues. The survey was designed with a top-down approach (Sutton 2015). Qualtrics was used as a survey tool that allowed researchers to create, distribute, and analyse responses from one convenient online location (Libguides 2011). The data generated from the various survey results was then analysed and presented in graph and table format, providing the reader with a visual representation of the current market feedback.

8.9.4 Data recording, management, and analysis methods

The study uses primary and secondary data sources to develop an analysis of the current construction industry. The primary data was collected from targeted members of the industry who demonstrate proficiency in BIM and digital engineering backgrounds. In order to examine the status of the industry, the research planned to obtain a significant amount of data from the respondents. The survey was completely anonymous and voluntary.

The responses were then analysed and categorised into general, technical, beneficial, integration, training, implementation, categorised, obstacle, and open-ended questions. The survey had 34 questions which were divided into open-ended and rank order questions. The research used a five-point Likert scale ranging from −2 "Strongly Disagree" (defined as not significantly important) to +2 "Strongly Agree" (defined as significantly important). A score of zero was given to respondents who selected neutral. Previous research undertaken by Kurwi, Demian et al. (2021) determined the most effective techniques used to analyse and display Likert data gathered from the survey were basic descriptive statistics, including measures of means and basic charts and graphs.

Secondary data collected from literature reviews was compared and quantified with primary data obtained from the respondents. The results were analysed against the current technologies implemented compared to previous arrangements. The responses were collected anonymously via the online Qualtrics survey platform by the first author. Literature reviews also collected secondary data from data sources such as journals, articles, and websites that had an interest in and knowledge of construction with BIM/GIS systems.

8.10 Findings

The research conducted into BIMs' impact on Australian Government projects tries to answer the problematic questions, including:

- Why have very slow progressive changes occurred, and how can momentum be gained in BIM implementation in the Australian Government construction sector?
- Are project and stakeholder communication inefficiencies a fundamental barrier?
- What strategies can be developed to improve the Australian BIM adoption model?
- Form a comparative review of the status of overseas models versus Australia.

The data was exported from the Qualtrics survey and analysed through a mixed-method approach (Kalu and Bwalya 2017). The quantitative data was collected from 37 of the 55 questionnaire responses. Of the 55 responses, 17 were incomplete, and four were invalid responses. The sample size is one of the study's limitations, which does not allow for generalisation of the investigation's findings.

8.10.1 General questions

The findings of the general questions concluded that the three highest numbers of survey participants were members of architects' engineering or consultancy (AEC) firms (Azhar 2011). The questions consisted of experience, gender, organisation size and geographical region, and the results are shown in Figures 8.1–8.4. BIM has been recognised as assisting AEC firms in planning, designing, constructing, and operating assets. This chapter also found that AEC firms were the largest users among the respondents. In fact, over 60% of the participants fit this criterion.

With AEC being the most frequent user, the data extracted from the primary sources was utilised for industry practitioners who have considered adopting BIM technology for their organisations. BIM has significant importance in the AEC industry since it enables the integration of the roles of the stakeholders project-wide (Azhar 2011). The gender of BIM specialists who took part in the psurvey was predominantly male, as shown in Figure 8.2.

Furthermore, it is also possible to conclude that most of the questionnaire participants from senior to middle management levels were also male. Of the 38 participant responses analysed, there were industry practitioners with up to 31 years of experience. More than 70% of the responses were from large organisations with >200 employees. The Australian construction industry is host to one of the top three largest businesses with the greatest increase. There are 410,839 construction businesses operating in Australia (Australian Bureau of Statistics 2021), 0.23%

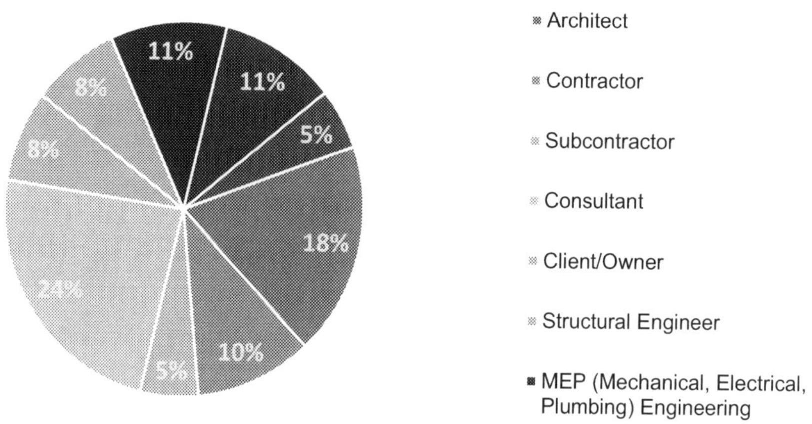

Figure 8.1 Respondent categories based on the sample data.

Source: Author's data.

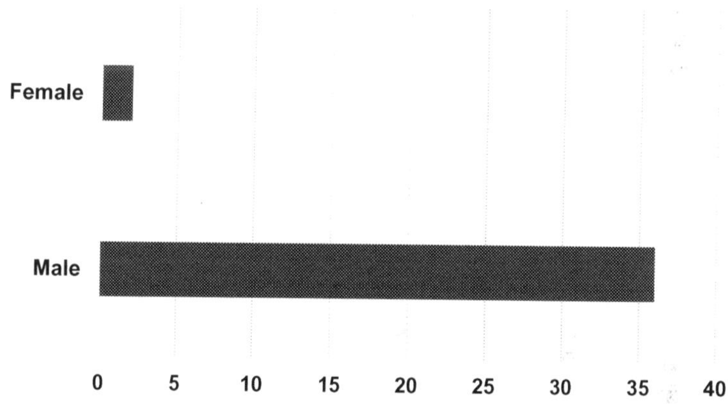

Figure 8.2 Gender demography of the participants.

(4,368) of which are large organisations. SMO businesses take up 99.7% of the construction sector. Both in the Australian industry and within this study, SMOs are categorised as organisations with less than 200 employees. Previous research concluded that SMOs are slow to adopt BIM when compared to larger pilots (Engineers Australia 2014). The research also indicates there are still significant doubts surrounding the return on investment and the cost of implementing BIM technologies. This analysis also discovered that the larger companies (>200 employees) were primarily based in New South Wales.

Figure 8.3 represents the annual per cent change in businesses by employment size in 2020–2021 (Australian Bureau of Statistics 2021). There has been a significant increase in SMOs. They have experienced over 80% increase in numbers. Large organisations increased by 15% in the same period. The graph in Figure 8.3

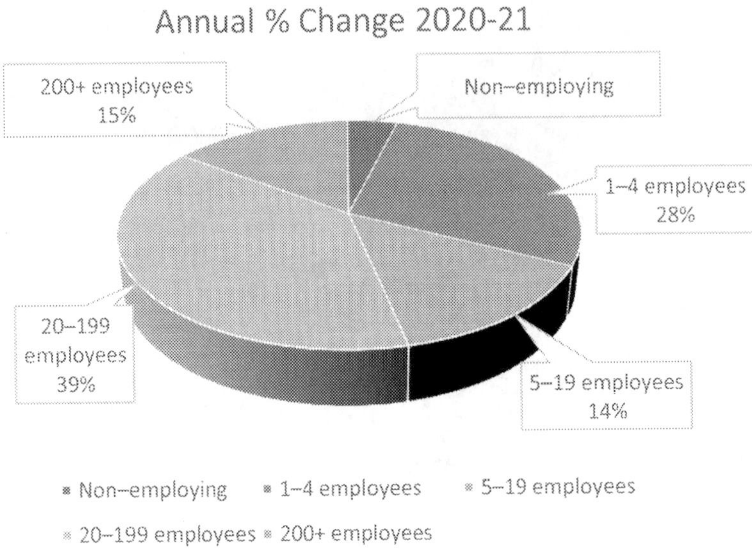

Figure 8.3 Annual employment changes based on the sample data.

indicates that SMOs are in high demand in the industry. The government needs to consider if BIM implementation switches from an industry-driven fragmented format to a government-led policy-driven one.

8.10.2 Integration

The impacts of BIM in Australian infrastructure projects have been mixed. The collection of responses concluded that many larger organisations were working nationwide. Table 8.2 shows project numbers and the size-range of companies delivering the projects located in different states in Australia.

The benefits of BIM integration have been examined in previous studies. Traditionally, design and construction phases were fully coordinated or integrated. However, the anticipated reduction in construction costs related to cost and programme and design collaboration has not been achieved (Luth, Schorer et al. 2014). Instead, BIM technology was being used in its basic form, namely for contract requirements or clash detection.

The results from the survey indicated that the top three construction projects being undertaken were rail, public sector, and commercial/industrial. In the context of Infrastructure Australia, the findings align with the national plan to improve and upgrade rail and infrastructure networks. As shown in Figure 8.4, the row labelled "others" consisted of Defence and Education, Utilities, Education, Interiors, Health, Mining Resources and Education. Almost all of the 38 responses utilised BIM-GIS integration on past or current projects. Of the four who had not used BIM-GIS integration, it was further analysed that BIM and GIS were used separately, for example, during site selection and ecological assessment phases.

Table 8.2 Employee size of various companies in Australia

No. of full-time staff at organisation					
Location of projects undertaken	*Total*	*Micro (0–4 employees)*	*Small (5–19 employees)*	*Medium (20–199 employees)*	*Large (>200 employees)*
NSW	36	3.0	2.0	4.0	27.0
VIC	20	2.0	1.0	2.0	15.0
QLD	20	2.0	1.0	2.0	15.0
ACT	16	0.0	0.0	2.0	14.0
NT	10	0.0	0.0	0.0	10.0
WA	19	3.0	0.0	1.0	15.0
SA	16	0.0	1.0	0.0	15.0
TAS	9	0.0	0.0	0.0	9.0

The findings from this category also conclude that architects' and multidisciplinary design managers' requirements for the use of digital twins in urban contexts for visualisation and the delivery of projects.

BIM technology has been used to assist with analysing green buildings and sustainability. The global interest surrounding climate change has placed building designers in the lens of environmentalists and building ecology specialists. By reducing the impact building projects have on the environment and improving buildings' overall performance, we can start to challenge ourselves to achieve a more ecological result (Luth, Schorer et al. 2014). To assist designers and improve effectiveness, BIM has been combined with optimised models to explore alternative avenues of design and has also been used as a key decision-making enabler (Asl, Zarrinmehr et al. 2015).

8.10.3 Potential benefits

The benefits of BIM-GIS technologies have become recognised within the planning strategies of government-led infrastructure projects. The results of the survey responses regarding benefits support GIS technologies with their greater data availability as a key decision enabler. The collaboration between stakeholders can be improved and the risk levels associated with the project can be greatly reduced by using the tools. There were split responses in relation to the statement, "GIS improves productivity and accuracy in quantity take off." Despite the mixed responses, the research data confirms that GIS improves productivity in a geographical context (Cheng and Yang 2001); the authors look at water productivity in spatial and temporal dimensions.

In promoting BIM in practice, when the industry was asked about the benefits of BIM under specific disciplines, there were more conflicting responses. Seventy-five per cent of the responses neither agreed nor disagreed and disagreed with BIMs' ability to indicate project overrun early. A further 94% neither agreed nor disagreed that BIM shortened the procurement schedules. This conflicts with the

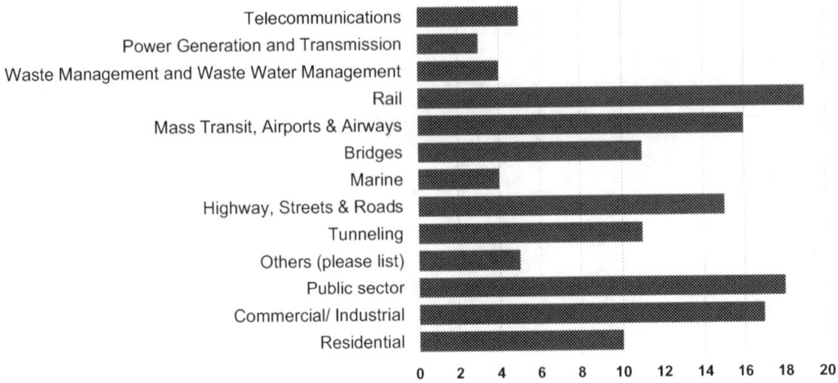

Figure 8.4 Project types of the selected sample.

literature reviewed by Aguiar Costa and Grilo (2015). Within this study, Aguiar Costa and Grilo (2015) propose how BIM-based solutions to support procurements may reduce the negative effects of the fragmentation of the building lifecycle process. There are clear discrepancies between this study and the questionnaire responses received in the current research and one can conclude that there are clear knowledge gaps in both sources.

Despite the inconsistencies, most of the stakeholders were neutral to very satisfied with BIM-GIS. Only one dissatisfied response was recorded. The benefits of BIM-GIS technology are evident in Figure 8.5. Between 70% and 74% of respondents agreed that BIM-GIS were both beneficial for successful project delivery and improved project visualisation. Despite there being clear evidence indicating the benefits of the integrated technology, 88.89% of the responses were uncertain of the outcomes of BIM-GIS integration. The future for BIM implementation looks challenging but achievable.

The United States has led the way in developing national BIM standards, while some European countries have reached between 60% and 70% application rate. Alternative industries, i.e., agriculture, aviation, marine, and space exploration, are understanding the value that BIM integration can bring to their projects, including holistic approaches to industry (Khan, Aziz et al. 2018; Mahdi, Mohamedien et al. 2019; Abbondati, Biancardo et al. 2020; Ma, Song et al. 2020).

However, the responses vary in the context of agriculture, marine science, aviation and the space domain. The current digital platforms in place for agriculture are "ahead of the construction industry." Advanced farm management software is driving digital crop growth whilst the water and river management systems are yet to be explored in Australia.

China has taken innovative steps towards integrating "BIM + GIS + VR." China has unlocked a first of its kind, the world's largest automatic monitoring network for the water environment. The automatic water quality monitoring station (AWQMS) consists of water quality analysis modules and multi-sensing water extraction and control systems (Li, Lu et al. 2021). The system was designed to

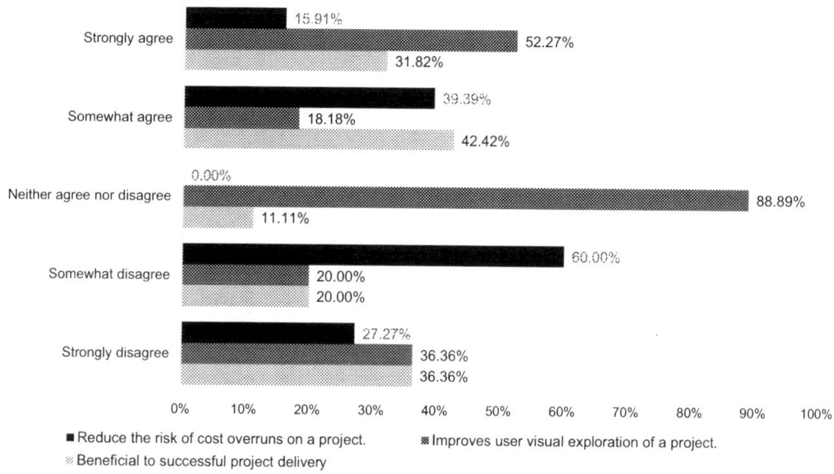

Figure 8.5 Participants' views on the benefits of BIM-GIS integration.

monitor and manage water usage in China. Fusing BIM and GIS technology with the virtual and real environment improves visualisation and future projects' spatial capability.

8.10.4 Implementation of BIM-GIS

The most noticeable outcomes occur when BIM-GIS is implemented at the conceptual design stages of a project. Both programs can be utilised in parallel for the duration of the project. This process is continuously reviewed and updated throughout the project lifecycle with specific deliverables for the conceptual and operational stages of the project. The responses from the implementation group of questions suggest that a large proportion of the industry agreed that it lacks a BIM standard, as shown in Figure 8.6. To create a vibrant and consistent BIM, it requires Government bodies to be involved. To date, there has been no mandate set by the Australian Government. In its place, there is a partial requirement for BIM to be included in all government projects greater than $50 million and a $30 million mandate for NSW Health Infrastructure. A government-led mandate would be a key initiative leading to the adoption and integration of Australian industry into the global network of opportunities. A mandate would also encourage critical research and training that is imperative to Australia's future workforce and BIM success.

As expressed previously, SMOs account for the majority of industry players. Despite the Australian national and state governments' plans to segregate larger contracts to enable SMOs to participate in large infrastructure projects, SMOs have the lowest BIM application rates. There is a growing case for considering the wider economic benefits of BIM on SOMs in infrastructure projects. The major concerns that exist with SMOs indicate the high cost of implementing the hardware

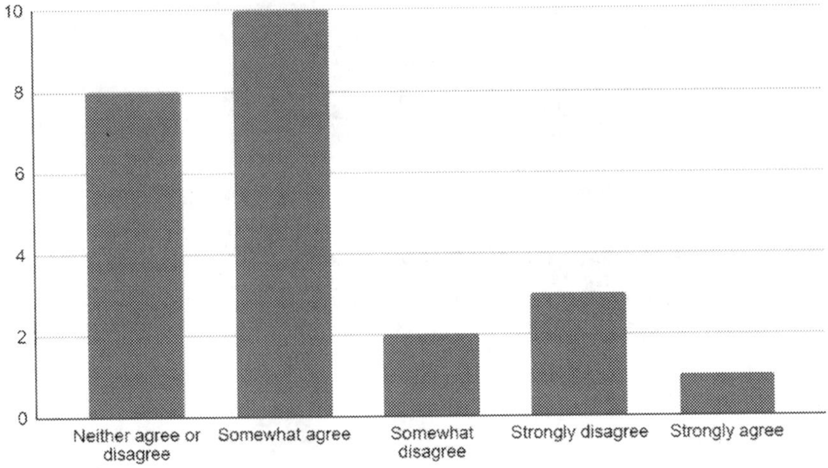

Figure 8.6 Lack of detailed BIM standards from the participants' view, based on the number of participants' agreement.

and maintenance investment and the uncertainty of investment returns post-BIM financial investment.

Given the opportunity, most users would use BIM-GIS technologies for future projects. However, there was some resistance, which indicates that unsatisfactory outcomes were experienced during its usage.

8.10.5 Technical issues

The Australian construction industry has not yet witnessed the capacity of full BIM integration. As previously mentioned, BIM technologies were traditionally used for clash detection or a contract requirement. Figure 8.7 shows that the responses indicate that over 90% of BIM users are utilising the technology for 3D modelling, which aligns with level 1 "2D/3D Partial Collaboration" of the BIM maturity levels graph.

BIM-GIS-VR implementation is the gateway to solving many unforeseen issues within the construction industry. Multidisciplinary industries are also heavily investing in augmented, virtual, and mixed reality technologies with the hope of creating a highly interactive sensory experience, engaging through touch, sound, and visual content (Australian Government 2020). The development of this technology is rapidly transforming how we plan and approach new projects across all industries. However, the accelerated rate of transformation also presents risks in relation to safety, privacy, and security. Furthermore, similar to BIM, no Australian VR standards have been developed.

During the questionnaire survey, users of these technologies pronounced significant issues that require further investigation. The problems embedded in the technologies need to be tackled to ensure long-term usability. Compatibility and transfer of data between model double-ups and an ecosystem of data with no

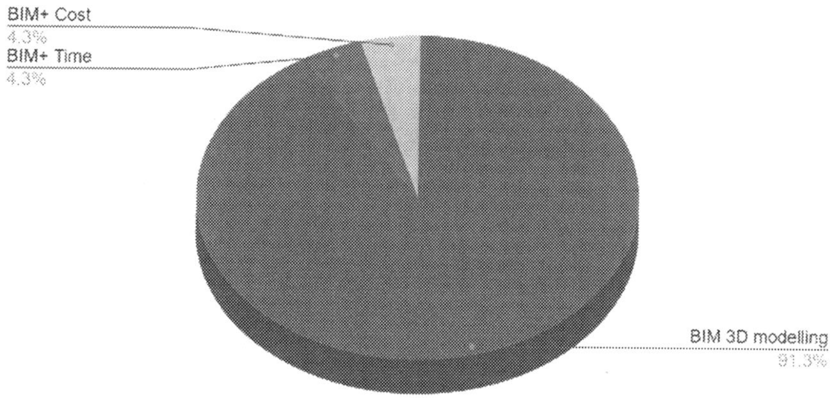

Figure 8.7 The common BIM applications based on the survey sample.

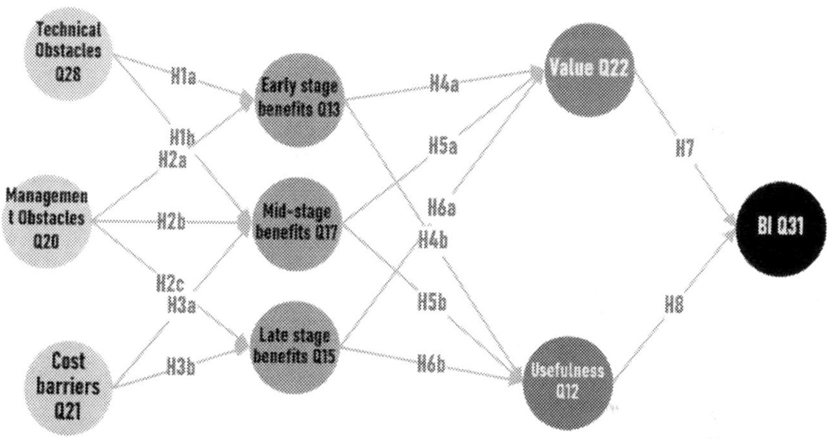

Figure 8.8 The conceptual model of BIM-GIS benefits and users' intention assessment, including 15 hypothetical paths and nine key constructs.

alignments are just two of the issues related. To avoid such issues and to ensure successful implementation, all platforms (BIM-GIS-VR) must share common coding and data standards.

8.10.6 The proposed adoption model

This section examines a proposed model of BIM-GIS benefits, illustrated in Figure 8.8, including various constructs listed in Table 8.3. The model was analysed using structural equation modelling, and the outcome discussed as follows.

In order to measure the reliability and internal consistency of the variables, Cronbach's alpha was computed, which displays a high level of convergent validity. The results of the test are shown in Table 8.4. Cronbach's alpha are all above

0.6, ranging from 0.815 to 1.00 (Cronbach 1951; Hair Jr, Hult et al. 2016). The composite reliability (CR) coefficient is also calculated to examine the correlation among variables and the fitness of variables to the model (Fornell and Larcker 1981). Table 8.4 shows that the CR coefficients vary from 0.889 to 1.00, which is acceptable as >0.6 (Fornell and Larcker 1981). Similarly, the AVE values are >0.5, so they are acceptable since they range between 0.608 and 1.00 (Fornell and Larcker 1981).

In order to test the independence and level of correlation, the Fornell and Larcker (1981) criteria were evaluated. Table 8.5 shows the discriminant validity matrix, which confirms the convergent and discriminant validities of the variables in the present chapter. In the matrix, the diagonal cells show a high correlation between the proposed factors and, varying from 0.780 to 1.000, well above the suggested values of 0.70.

Table 8.6 shows the results of various tests, including f^2, with a value ranging from 0.003 to 0.618 for every path. This value for the supported paths is >0.02, which is acceptable, according to Hair Jr, Hult et al. (2016). The table also shows the path coefficients and t-statistics of every path and shows what hypothetical paths are supported.

8.10.7 Limitations

This chapter provides a theoretical model that can potentially be a predictor of interactive, immersive tools. However, the model needs to be tested by different groups of participants from different regions to be able to generalise it. Similar to other surveys, the outcome of one survey cannot be generalised; a discussion of the theoretical model is needed as a prelude to possible modifications and testing. That will help scholars evolve the theoretical concept behind the model to be applied to larger empirical studies. Future studies need to discuss more detailed performance expectations regarding immersive interactive tools and establish how practitioners can use them in construction activities such as building façades, excavating, pouring concrete, and using tower cranes.

8.11 Discussion

Despite the apparent contradictions between the findings within this research and the literature reviews, this research concludes BIM holds significant value within the construction industry, and in the context of larger construction projects, costs can be vastly improved. However, it would take much longer for SMOs to experience the uplift of BIM benefits. With the aid of new technologies available, namely, VR, AR, and MR, the quality of construction will certainly be improved by applying BIM-GIS technologies. Collaborative contracting, although in its infancy, has been established and governments are leading the way on infrastructure projects. However, there are not many updated generalised guidelines for alternative projects. The development of the National BIM Guide in 2011 (NATSPEC 2021) was established to standardise BIM practices in the previous years.

Table 8.3 A set of descriptive statistics including the significance values, loading, and VIF

ID	*Constructs*	*Original sample (O)*	*Sample mean (M)*	*Standard deviation*	*T statistics*	*P values*	*Loading*	*Outer VIF*
BI	Behavioural intention	1.000	1.000	0.000			1.000	1.000
CO2	Cost barriers	0.832	0.795	0.159	5.239	0.000	0.832	1.564
CO3		0.843	0.793	0.184	4.577	0.000	0.843	2.080
CO4		0.883	0.864	0.126	7.019	0.000	0.883	2.009
ESB1	Early stage benefits	0.833	0.822	0.085	9.743	0.000	0.833	3.283
ESB2		0.843	0.843	0.066	12.699	0.000	0.843	3.330
ESB3		0.926	0.910	0.078	11.828	0.000	0.926	4.867
ESB4		0.919	0.906	0.080	11.546	0.000	0.919	4.863
ESB5		0.844	0.825	0.100	8.428	0.000	0.844	2.593
LSB1	Late stage benefit	0.979	0.975	0.018	54.916	0.000	0.979	10.564
LSB2		0.945	0.931	0.058	16.343	0.000	0.945	6.486
LSB3		0.902	0.903	0.036	24.752	0.000	0.902	3.680
MO1	Mid-stage benefit	0.880	0.876	0.066	13.351	0.000	0.880	2.209
MO2		0.905	0.896	0.069	13.208	0.000	0.905	2.799
MO3		0.898	0.885	0.102	8.766	0.000	0.898	2.320
MSB1	Management obstacles	0.855	0.847	0.053	16.049	0.000	0.855	3.837
MSB2		0.892	0.883	0.059	15.101	0.000	0.892	4.309
MSB3		0.853	0.846	0.062	13.829	0.000	0.853	4.745
MSB4		0.848	0.842	0.056	15.160	0.000	0.848	3.828
MSB5		0.880	0.862	0.085	10.302	0.000	0.880	5.552
MSB6		0.947	0.941	0.032	29.859	0.000	0.947	11.731
MSB7		0.914	0.907	0.048	18.967	0.000	0.914	6.888
MSB8		0.925	0.919	0.036	25.414	0.000	0.925	7.942
TO1	Technical obstacles	0.893	0.871	0.129	6.935	0.000	0.893	2.198
TO2		0.804	0.787	0.151	5.319	0.000	0.804	1.572
TO4		0.880	0.854	0.109	8.089	0.000	0.880	2.500
USF1	Usefulness	0.910	0.903	0.070	13.094	0.000	0.910	9.782
USF11		0.735	0.726	0.100	7.381	0.000	0.735	2.382
USF2		0.895	0.891	0.066	13.513	0.000	0.895	8.954
USF3		0.906	0.893	0.074	12.201	0.000	0.906	4.883
USF4		0.767	0.749	0.121	6.362	0.000	0.767	2.105
USF5		0.897	0.889	0.066	13.538	0.000	0.897	4.266
USF7		0.779	0.775	0.079	9.882	0.000	0.779	2.475
VAL2	Values	0.781	0.776	0.094	8.325	0.000	0.781	2.967
VAL3		0.725	0.722	0.124	5.826	0.000	0.725	2.011
VAL4		0.810	0.802	0.095	8.489	0.000	0.810	2.457
VAL5		0.730	0.696	0.172	4.249	0.000	0.730	3.744
VAL6		0.762	0.744	0.136	5.614	0.000	0.762	2.965
VAL7		0.864	0.849	0.095	9.119	0.000	0.864	3.652

Table 8.4 The outcome of the tests confirms the level of reliability and validity required for modelling

Factors	*Cronbach's alpha*	*rho_A*	*Composite reliability*	*Average variance extracted (AVE)*	*Q² (=1 – SSE/SSO)*	*R squared*	*R squared adjusted*
BI	1.000	1.000	1.000	1.000	0.512	0.598	0.574
CO	0.815	0.836	0.889	0.727		0.479	0.448
ESB	0.923	0.935	0.942	0.764	0.322	0.294	0.252
LSB	0.937	0.948	0.960	0.889	0.230	0.334	0.273
MO	0.875	0.882	0.923	0.800		0.469	0.421
MSB	0.962	0.964	0.968	0.792	0.243	0.275	0.209
TO	0.825	0.855	0.895	0.740		0.598	0.574
USF	0.931	0.935	0.945	0.713	0.304	0.479	0.448
VAL	0.871	0.881	0.903	0.608	0.140	0.294	0.252

Table 8.5 Confirmation of discriminant validity in this study based on Fornell and Larcker's (1981) criteria

	BI	*CO*	*ESB*	*LSB*	*MO*	*MSB*	*TO*	*USF*	*VAL*
BI	1.000								
CO	−0.550	0.853							
ESB	0.745	−0.384	0.874						
LSB	0.652	−0.431	0.735	0.943					
MO	−0.596	0.506	−0.634	−0.502	0.894				
MSB	0.683	−0.292	0.823	0.655	−0.560	0.890			
TO	−0.689	0.605	−0.653	−0.629	0.732	−0.500	0.860		
USF	0.689	−0.437	0.669	0.586	−0.641	0.590	−0.696	0.844	
VAL	0.591	−0.477	0.510	0.323	−0.435	0.470	−0.497	0.389	0.780

There are, however, significantly low BIM adoption rates in SMOs. SMOs are accountable for close to 98% of companies in the construction industry. Of the 98%, 70% are non-adopters. There is an intrinsic gap between large and SMOs in the Australian construction industry, and mandating BIM would pose future challenges to the return on investment for smaller organisations. The barrier will remain without a specific Australian standard incorporating BIM processes or government incentives to assist traditional and encourage new organisations. Although the International Organization for Standardisation (ISO) has developed ISO 19650, BIM standards look at BIM in a broad way, focusing on the legal and contractual impacts of the digitisation of information management.

Training and education are fundamental in developing BIM users' skill sets. However, the training provided has been moving slowly, which led to the creation of CodeBIM. Collaborative BIM (CodeBIM) was designed using BIM technologies to encourage collaboration between AEC students and to reshape AEC courses and programs to meet the industry's needs. Higher education institutions play an important role in delivering a broad industry awareness; however, solely depending on educational institutions to deliver BIM training is both short-sighted and impractical. An industry and education combined effort is required.

The US has paved the way for developing national BIM standards. Mega infrastructure projects have seen value in BIM. The annual applications are also increasing in industries, including green buildings, tunnelling, urban planning, and environmental protection. To tackle the widespread problem of urbanisation, the Australian Government and state authorities must continuously upgrade the infrastructure system and provide a solution to urban regions around the country. Despite attempts to capitalise on integrating BIM services with existing digital models, worldwide construction industry practice is slow to embrace change. BIM

Table 8.6 The outcome of model testing includes statistical values, inner VIF, effect size, and the supported hypotheses

Path	*Original sample (O)*	*Sample mean (M)*	*Standard deviation (STDEV)*	*T statistics (|O/STDEV|)*	*P values*	*Inner VIF*	f^2	*Hypothesis acceptability*
CO → LSB	−0.238	−0.237	0.201	1.183	0.237	1.344	0.060	Not supported
CO → MSB	0.058	0.073	0.206	0.284	0.777	1.598	0.003	Not supported
ESB → USF	0.447	0.311	0.321	1.394	0.163	3.925	0.096	Not supported
ESB → VAL	0.461	0.395	0.348	1.324	0.185	3.925	0.075	Not supported
LSB → USF	0.196	0.213	0.237	0.829	0.407	2.214	0.033	Not supported
LSB → VAL	−0.131	−0.099	0.253	0.518	0.604	2.214	0.011	Not supported
MO → ESB	−0.334	−0.310	0.148	2.266	0.024	2.156	0.099	Supported
MO → LSB	−0.381	−0.361	0.166	2.292	0.022	1.344	0.153	Supported
MO → MSB	−0.427	−0.418	0.212	2.007	0.045	2.186	0.125	Supported
MSB → USF	0.094	0.192	0.357	0.263	0.793	3.155	0.005	Not supported
MSB → VAL	0.176	0.164	0.286	0.616	0.538	3.155	0.014	Not supported
TO → ESB	−0.408	−0.406	0.196	2.084	0.037	2.156	0.148	Supported
TO → MSB	−0.223	−0.221	0.232	0.961	0.337	2.565	0.029	Not supported
USF → BI	0.541	0.520	0.110	4.923	0.000	1.178	0.618	Supported
VAL → BI	0.380	0.361	0.132	2.887	0.004	1.178	0.305	Supported

is now considered part of the Australian construction industry and is predominantly used on larger, complex projects. Therefore, plans are being made for its wider adoption; however, the lack of skilled expertise to fill these roles remains.

The study is limited to the sample and the government-funded projects in Australia. During the research, Australian levels of BIM technology were determined to be at infancy levels when compared with overseas practitioners, i.e., Finland, North America, and the United Kingdom (Johnston 2008). This also aligns with the primary data extracted from the survey responses. The research also found that the Australian industry version of BIM modellers had previous roles as CAD designers. Although there is a relationship between both software systems, the relationship only exists at the infancy level, namely BIM level 0. Therefore, the level of professionals targeted during the survey is limited to initial levels of BIM expertise within Australia.

According to reports conducted by the Institution of Public Works Engineering Australasia (Institute of Public Works Engineering Australasia 2019b), BIM implementation differs between each state, which raises some confusion with the lack of consistency. The lack of consistency towards the technology surrounding BIM and with BIM itself has also raised resistance towards implementation. This misalignment adds to the disjointed level of state respondents to the survey (De Cicco 2021).

The lack of digital standards, national mandate, and slow-paced technology do not support manufacturers in Australia. Further investigation of BIM-GIS would demonstrate the need for strategic vision to compete at the cutting edge of technology. Finally, it is recommended that future research be given a higher priority to responding to verified respondents. In addition, the timeframe should be extended to a few years as a longitudinal survey to provide more information about the users' behavioural trends.

8.12 Conclusions

BIM is a complete project communication and collaboration package. The implications of this study suggest the BIM approach offers multiple benefits to all project stakeholders involved, including members of the AEC industry and urban and infrastructure planners. By including all disciplines, stakeholders can truly reap the rewards from BIM. It has been proven that companies' productivity can be increased when BIM is fully implemented. Using the platform can contribute to greater success and fundamentally better outcomes. The Australian Government may need to lead a more collaborative approach toward BIM implementation; a proprietary regulatory system seems to be the most effective route. Instead of only facilitating the apex players of the industry, i.e., the Tier 1 contractors, the government needs to be at the forefront of a new approach to ensure we have a fully functional product and a regulatory system that reinforces it.

The contribution of the chapter lies in identifying the value of technical obstacles (TO) in the early stages of BIM-GIS utilisation at a company ($t = 2.084$), while cost obstacles (CO) are a matter for the later stages of the adoption ($t = 1.183$). The study also identified that the managerial obstacles (MO) would affect all stages ($t >$

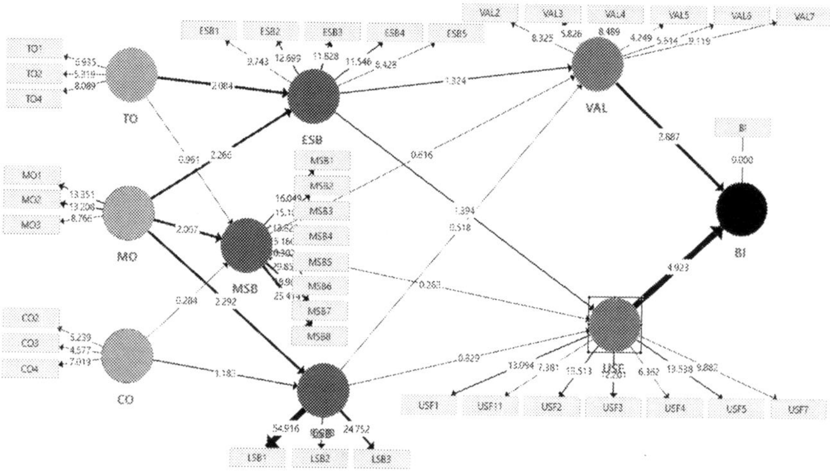

Figure 8.9 The schematic model of the BIM-GIS values and intention to use tested, including the relative values of path coefficients and *t*-values.

2.007), from early to late, of technology adoption in government-funded infrastructure projects.

The chapter also contributed to the body of knowledge in technology adoption modelling by identifying two constructs affecting users' intentions to use BIM and GIS, including values and usefulness for infrastructure projects. Value is measured by improvement of construction quality ($t = 8.325$), speeding up construction progress ($t = 5.826$), increasing safety in the design stage ($t = 8.489$), faster and easier calculation of the workflows ($t = 4.249$), dramatically enhanced level of accuracy during the prefabrication of materials ($t = 5.614$, refer to Figure 8.9), and general site safety attributes ($t = 9.119$, refer to Figure 8.9). The chapter filled the gap in investigating the adoption factors of both information modelling systems' BIM and GIS' in the infrastructure sector, where BIM and GIS are required for information modelling. Despite changes we have seen in favour of the digital industry, small changes are not the solution. Research and training are improving, but the rate is slow and stagnant. If the construction industry is to progress, it requires rapid and large-scale action.

References

Abbondati, F., S. A. Biancardo, S. Palazzo, F. S. Capaldo and N. Viscione (2020). "I-BIM for existing airport infrastructures." *Transportation Research Procedia* **45**: 596–603.

Aguiar Costa, A. and A. Grilo (2015). "BIM-based e-procurement: An innovative approach to construction e-procurement." *The Scientific World Journal*. **2015**: 905390. doi: 10.1155/2015/905390

Asl, M. R., S. Zarrinmehr, M. Bergin and W. Yan (2015). "BPOpt: A framework for BIM-based performance optimization." *Energy and Buildings* **108**: 401–412.

Australasia, b. (2012). National Building Information Modelling Initiative. Vol. 1, Strategy: A strategy for the focussed adoption of building information modelling and related digital technologies and processes for the Australian built environment sector, report to the Department of Industry, Innovation, Science, Research and Tertiary Education, Sydney, June. https://www.pc.gov.au/inquiries/completed/infrastructure/submissions/submissions-test2/submission-counter/subdr170-infrastructure-attachment.pdf.

Australian Bureau of Statistics (2019). *Measuring digital activities in the Australian economy*. A. B. o. Statistics.

Australian Bureau of Statistics (2021). *Counts of Australian businesses, including entries and exits*.

Australian Government (2015). *National framework for traditional contracting the guide good practice and commercial principles for traditional contracting*. I. a. R. Development, Department of Infrastructure and Regional Development.

Australian Government (2018). *Industry insights: 3/2018 future productivity*.

Australian Government (2020). *Immersive technologies – Position statement*. e. Commissioner.

Azhar, S. (2011). "Building information modeling (BIM): Trends, benefits, risks, and challenges for the AEC industry." *Leadership and Management in Engineering* **11**(3): 241–252.

Back to Basics (2019). "Construction industry facts (updated 2020) - Back to Basics." Retrieved 28/07/2021, 2021, from https://backtobasics.edu.au/2019/03/construction-industry-facts/.

Barlish, K. and K. Sullivan (2012). "How to measure the benefits of BIM—A case study approach." *Automation in Construction* **24**: 149–159.

Bridge, C. and P. Carnemolla (2014). "An enabling BIM block library: an online repository to facilitate social inclusion in Australia." *Construction Innovation* **14**(4): 477–492.

Bris, A., C. Cabolis and J. Caballero (2017). "The IMD world digital competitiveness ranking."

Cardeira, H. (2019). "A brave new form of processing payments in the construction industry: An investigation into how distributed ledger technology can address the security of payment problem in the construction industry."

Cheng, M.-Y. and S.-C. Yang (2001). "GIS-based cost estimates integrating with material layout planning." *Journal of Construction Engineering and Management* **127**(4): 291–299.

Cronbach, L. J. (1951). "Coefficient alpha and the internal structure of tests." *Psychometrika* **16**(3): 297–334.

De Cicco, R. (2021). "NBS BIM object standard." https://www.thenbs.com/our-tools/nbs-bim-object-standard

Department of Industry, Science, Energy and Resources (2018). https://www.industry.gov.au/publications/australia-2030-prosperity-through-innovation

Deshpande, A., S. Azhar and S. Amireddy (2014). "A framework for a BIM-based knowledge management system." *Procedia Engineering* **85**: 113–122.

Dowsett, R. and C. Harty (2013). *Evaluating the benefits of BIM for sustainable design–A review*. In 29th Annual ARCOM Conference, Smith, S.D and Ahiaga-Dagbui, D.D. Association of Researchers in Construction Management: 13–23.

Dyson, L. (2021). "Digital revolution will transform UK national highways' roads." Retrieved 07/11/2021, 2021, from https://www.traffictechnologytoday.com/news/data/digital-revolution-will-transform-uk-national-highways-roads.html#:~:text=%E2%80%9CNational%20Highways'%20vision%20for%20Digital,chair%20of%20ITS%20(UK).

Eastman, C. M. (2008). *BIM handbook: A guide to building information modeling for owners, managers, designers, engineers, and contractors*, John Wiley & Sons.

Elghaish, F., S. Matarneh, S. Talebi, M. Kagioglou, M. R. Hosseini and S. Abrishami (2021). "Toward digitalization in the construction industry with immersive and drones technologies: A critical literature review." *Smart and Sustainable Built Environment* **10**(3): 345–363.

Elmualim, A. and J. Gilder (2014). "BIM: Innovation in design management, influence and challenges of implementation." *Architectural Engineering and Design Management* **10**(3–4): 183–199.

Engineers Australia (2014). "Welcome to Engineers Australia Portal." Retrieved 03/08/2021, 2021, from https://portal.engineersaustralia.org.au/news/driving-building-information-modelling-bim-uptake.

Esri (2021). "What is GIS?" Retrieved 25/11/2021, 2021, from https://www.esri.com/en-us/what-is-gis/overview.

Fischer, M. (2016). *Building information modeling can transform how we create cities*, Stanford University.

Fornell, C. and D. F. Larcker (1981). "Evaluating structural equation models with unobservable variables and measurement error." *Journal of Marketing Research* **18**(1): 39–50.

Forsythe, P. (2014). *The case for BIM uptake among small construction contracting businesses*. ISARC. Proceedings of the International Symposium on Automation and Robotics in Construction, CiteSeer.

Forsythe, P., S. Sankaran and C. Biesenthal (2015). "How far can BIM reduce information asymmetry in the Australian construction context?" *Project Management Journal* **46**(3): 75–87.

Gerrard, A., J. Zuo, G. Zillante and M. Skitmore (2010). Building information modeling in the Australian architecture engineering and construction industry. *Handbook of research on building information modeling and construction informatics: Concepts and technologies*, Jason Underwood: 521–545.

Ghaffarianhoseini, A., J. Tookey, A. Ghaffarianhoseini, N. Naismith, S. Azhar, O. Efimova and K. Raahemifar (2017). "Building Information Modelling (BIM) uptake: Clear benefits, understanding its implementation, risks and challenges." *Renewable and Sustainable Energy Reviews* **75**: 1046–1053.

Griffey, S. (2019). "Future of BIM in Australia." Retrieved 21/06/2021, 2021, from https://morrisseylaw.com.au/future-of-bim-in-australia/.

Hair Jr, J. F., G. T. M. Hult, C. Ringle and M. Sarstedt (2016). *A primer on partial least squares structural equation modeling (PLS-SEM)*, Sage Publications.

Hamid, T. (2021). "The evolution of digital engineering and its impact on project delivery." Retrieved 07/11/2021, 2021, from https://roadsonline.com.au/the-evolution-of-digital-engineering-and-its-impact-on-project-delivery/.

Highways England (2014). "Digital roads." Retrieved 07/11/2021, 2021, from https://nationalhighways.co.uk/our-work/digital-data-and-technology/digital-roads/.

Infrastructure Australia (2016). *National infrastructure summit 2016*. Infrastructure Australia. Infrastructureaustralia.gov.au.

Infrastructure Australia (2019). *Australian infrastructure audit 2019*. Infrastructure Australia.

Institute of Management Development (2020). "World digital competitiveness rankings-IMD." Retrieved 28/07/2021 2021, from https://www.imd.org/centers/world-competitiveness-center/rankings/world-digital-competitiveness/.

Institute of Public Works Engineering Australasia (2019a). "What you need to know about BIM in Australia." Retrieved 05/08/2021, 2021, from https://www.ipwea.org/blogs/intouch/2016/08/01/what-you-need-to-know-about-bim-in-australia.

Institut of Public Works Engineering Australasia (2019b). "How BIM is slowly shaping Australia's infrastructure." Retrieved 03/08/2021, 2021, from https://www.ipwea.org/blogs/intouch/2018/07/25/how-bim-is-slowly-shaping-australias-infrastructur.

Johnston, R. (2008). "Book review: Mead, WR 2007: Adopting Finland. Helsinki: Hakapaino Oy. 151 pp.[Price?] ISBN: 978 952 92 1334 4 paper." *Progress in Human Geography* **32**(6): 852–854.

Kalu, F. A. and J. C. Bwalya (2017). "What makes qualitative research good research? An exploratory analysis of critical elements." *International Journal of Social Science Research* **5**(2): 43–56.

Kemp, D. (2019). "Sydney light rail legal disputes settled with revised PPP." Retrieved 27/05/2021, 2019, from https://www.infrastructureinvestor.com/sydney-light-rail-legal-disputes-settled-revised-ppp/.

Khan, R., Z. Aziz and V. Ahmed (2018). "Building integrated agriculture information modelling (BIAIM): An integrated approach towards urban agriculture." *Sustainable Cities and Society* **37**: 594–607.

Khosrowshahi, F. and Y. Arayici (2012). "Roadmap for implementation of BIM in the UK construction industry." *Engineering, Construction and Architectural Management* **19**(6): 610–635.

Kurwi, S., P. Demian, K. B. Blay and T. M. Hassan (2021). "Collaboration through integrated BIM and GIS for the design process in rail projects: Formalising the requirements." *Infrastructures* **6**(4): 52.

Leonhard (2016). "Digital transformation: Are you ready for exponential change?" Youtube, Futurist Keynote Speaker.

Li, T., Y. Lu, F. Chen, H. Xiao, L. Meng and Y. Hu (2021). *Application of BIM, VR and GIS technologies in water environmental in-situ monitoring and management*. International Conference on Smart Transportation and City Engineering 2021, SPIE.

Libguides (2011). "Research guides: Qualtrics: What is qualtrics?" Retrieved 04/08/2021, 2021, from //csulb.libguides.com/qualtrics/about.

Logan, V. (2018). "Enabling data literacy for digital society." *Information as a Second* Language: *Enabling Data Literacy for Digital Society*. Retrieved 07/11/2021, 2021, from https://www.gartner.com/en/documents/3602517.

Lu, W., Y. Peng, Q. Shen and H. Li (2013). "Generic model for measuring benefits of BIM as a learning tool in construction tasks." *Journal of Construction Engineering and Management* **139**(2): 195–203.

Luo, S., N. Yimamu, Y. Li, H. Wu, M. Irfan and Y. Hao (2023). "Digitalization and sustainable development: How could digital economy development improve green innovation in China?" *Business Strategy and the Environment* **32**(4): 1847–1871.

Luth, G. P., A. Schorer and Y. Turkan (2014). "Lessons from using BIM to increase design-construction integration." *Practice Periodical on Structural Design and Construction* **19**(1): 103–110.

Ma, G., X. Song and S. Shang (2020). "BIM-based space management system for operation and maintenance phase in educational office buildings." *Journal of Civil Engineering and Management* **26**(1): 29–42.

Mahdi, I., M. Mohamedien, H. Ibrahim and M. Khalil (2019). "Proposed management system of marine works based on bim approach (technology)." *Journal of Engineering and Applied Science* **66**: 771–790.

Marsh, P. (2021). "Vision set out for "digital revolution that will fundamentally change UK motorways"." Envirotec. Retrieved 07/11/2021, 2021, from https://envirotecmagazine.

com/2021/09/02/vision-set-out-for-digital-revolution-that-will-fundamentally-change-uk-motorways/.

NATSPEC (2021). "NATSPEC national BIM guide." Retrieved 26/11/2021, 2021, from https://bim.natspec.org/documents/natspec-national-bim-guide.

NSW Government (2021). *NSW infrastructure strategy 2018-2038*. Infrastructure NSW.

Oyewole, E. O. and J. O. Dada (2019). "Training gaps in the adoption of building information modelling by Nigerian construction professionals." *Built Environment Project and Asset Management* **9**(3):399–411.

Queensland Government (2021). *Building Information Modelling (BIM) for transport and main roads*. D. o. T. a. M. Roads.

Queensland Government Enterprise Architecture (2020). "BIM projects - Data and information guideline." Retrieved 21/06/2021, 2021, from https://www.forgov.qld.gov.au/information-and-communication-technology/qgea-policies-standards-and-guidelines/bim-projects-data-and-information-guideline.

Ribeirinho, M. J., J. Mischke, G. Strube, E. Sjödin, J. L. Blanco, R. Palter, J. Biörck, D. Rockhill and T. Andersson (2020). *The next normal in construction: How disruption is reshaping the world's largest ecosystem*. McKinsey & Company.

Sage Research Methods (2012). "Primary data source." https://methods.sagepub.com/reference/encyc-of-research-design/n333.xml.

Saranya, A. and R. Subhashini (2023). "A systematic review of Explainable Artificial Intelligence models and applications: Recent developments and future trends." *Decision Analytics Journal* **7**: 100230.

Sepasgozar, S. M. E., F. K. P. Hui, S. Shirowzhan, M. Foroozanfar, L. Yang and L. Aye (2021). "Lean practices using Building Information Modeling (BIM) and digital twinning for sustainable construction." *Sustainability* **13**(1): 161.

Sepasgozar, S. M., A. A. Khan, K. Smith, J. G. Romero, X. Shen, S. Shirowzhan, H. Li and F. Tahmasebinia (2023). "BIM and digital twin for developing convergence technologies as future of digital construction." *Buildings* **13**(2): 441.

Seyfarth (2019). "The integrated project delivery model: Why, what, and how to decide if it is right for your project." Retrieved 05/08/2021, 2021, from https://www.constructionseyt.com/2019/06/the-integrated-project-delivery-model-why-what-and-how-to-decide-if-it-is-right-for-your-project/.

Shepherd, A. (2021). "Victorian transport infrastructure conference." Retrieved 27/07/2021, 2021, from https://infrastructuremagazine.com.au/2021/06/23/nsw-budgets-108-5-billion-infrastructure-commitment/.

Sutton, I. (2015). *Process risk and reliability management*, 2nd ed., Gulf Professional Publishing.

Terreno, S., C. Anumba, E. Gannon and C. Dubler (2015). The benefits of BIM integration with facilities management: A preliminary case study. *Computing in Civil Engineering* **2015**: 675–683.

The University of British Columbia (2021). "Research." BIM Topics. Retrieved 08/11/2021, 2021, from https://www.united-bim.com/bim-maturity-levels-explained-level-0-1-2-3.

United-BIM Inc. (2020). "BIM maturity levels explained - Level 0| 1|2|3." Retrieved 05/08/2021, 2021, from https://www.united-bim.com/bim-maturity-levels-explained-level-0-1-2-3/.

Vaux, S. (2021). "Demystifying the digital asset lifecycle: The what, why, and how." www.deosdigital.com.

Visner, M., S. Shirowzhan and C. Pettit (2021). "Spatial analysis, interactive visualisation and GIS-based dashboard for monitoring spatio-temporal changes of hotspots of bushfires over 100 years in New South Wales, Australia." *Buildings* **11**(2): 37.

Yang, J.-B. and H.-Y. Chou (2018). "Mixed approach to government BIM implementation policy: An empirical study of Taiwan." *Journal of Building Engineering* **20**: 337–343.
Zhu, Z., S. Shirowzhan and C. J. Pettit (2022). "Investigation of development applications: A GIS based spatiotemporal analysis in the city of Sydney area 2004–2022." *Buildings* **12**(10): 1601.
ZIGURAT Global Institute of Technology (2018). "Lean BIM construction: Benefits of BIM and lean management." Retrieved 27/11/2021, 2021, from https://www.e-zigurat.com/blog/en/lean-bim-construction-benefits-of-bim-and-lean-management/.

Index

Note: **Bold** page numbers refer to tables and *italic* page numbers refer to figures.

Printed in the United States
by Baker & Taylor Publisher Services